KUMON MATH WORKBOOKS

Grades **6-8**

Algebra
Workbook I

Table of Contents

KUM N

Fractions Review

1

Level ☆

Date / /

Name

Score /100

1 **Calculate.**

5 points per question

(1) $0.5 + \dfrac{1}{9} =$

(6) $1\dfrac{3}{8} - 1.3 =$

 Rewrite decimals as fractions to calculate.

(2) $0.6 - \dfrac{1}{4} =$

(7) $0.75 - 0.25 + \dfrac{1}{2} =$

(3) $\dfrac{2}{5} + 0.75 =$

(8) $4.5 + \dfrac{1}{4} - 1.125 =$

(4) $1.75 - \dfrac{1}{6} =$

(9) $3\dfrac{1}{2} - 0.375 + 5\dfrac{1}{6} =$

(5) $\dfrac{3}{10} + 2.125 =$

(10) $9.25 - \dfrac{1}{2} + \dfrac{3}{8} =$

2 **Calculate.**

(1) $0.125 \times \dfrac{1}{4} =$

(6) $2.25 \div 1.5 =$

(2) $0.75 \div \dfrac{6}{11} =$

(7) $\dfrac{1}{3} \times 2.7 \times \dfrac{4}{5} =$

Remember to reduce as you calculate.

(3) $\dfrac{2}{3} \times 0.6 =$

(8) $\dfrac{3}{10} \div 1.4 \div \dfrac{3}{4} =$

(4) $\dfrac{1}{2} \times 3.5 =$

(9) $2.4 \times \dfrac{5}{6} \div 10 =$

(5) $0.375 \times 6 =$

(10) $1.8 \div 0.375 \times 1.25 =$

You're off to a great start!

1 **Calculate.**

(1) $2^5 \times \dfrac{1}{2^2} =$

(6) $2^2 \times 2^1 \times 2^2 = 2^{\square}$

$=$

> When multiplying exponents with the same base, add the exponents.

(2) $\left(\dfrac{3}{10}\right)^2 \times 2^4 =$

(7) $\left(\dfrac{2}{3}\right)^3 \times \left(\dfrac{2}{3}\right)^1 =$

> 2^1 is the same as 2.

(3) $\left(\dfrac{4}{9}\right)^2 \times \left(\dfrac{3}{8}\right)^2 =$

(8) $\dfrac{3^2}{4^1} \times \dfrac{3^0}{4^2} =$

(4) $\left(1\dfrac{1}{2}\right)^3 \times \left(\dfrac{8}{15}\right)^2 =$

(9) $\dfrac{2^3}{3^1} \times \dfrac{2^2}{3^1} \times \dfrac{2^1}{3^2} =$

> Any number raised to the 0 power equals 1.

(5) $\left(2\dfrac{1}{4}\right)^4 \times \left(1\dfrac{1}{3}\right)^3 =$

(10) $\dfrac{2^3}{5^1} \times \dfrac{3^0}{4^1} \times \dfrac{2^2}{5^0} \times \dfrac{3^2}{4^1} =$

2 **Calculate.**

5 points per question

(1) $2^6 \div 2^2 = 2^{\square} =$

(6) $\left(\dfrac{5}{2}\right)^2 \div (15)^2 =$

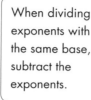

When dividing exponents with the same base, subtract the exponents.

(2) $3^{10} \div 3^4 \div 3^1 =$

(7) $\left(2\dfrac{1}{4}\right)^3 \div \left(3\dfrac{3}{4}\right)^3 =$

(3) $\left(\dfrac{2}{3}\right)^5 \div \left(\dfrac{2}{3}\right)^1 =$

(8) $\left(\dfrac{1}{2}\right)^2 \div \left(\dfrac{1}{2}\right)^1 \times \left(\dfrac{1}{2}\right)^3 =$

(4) $10^3 \div 2^3 =$

(9) $\left(\dfrac{3}{5}\right)^3 \times 6^3 \div \left(\dfrac{9}{10}\right)^3 =$

(5) $9^3 \div 12^3 =$

(10) $\left(1\dfrac{1}{3}\right)^2 \div \left(1\dfrac{2}{3}\right)^2 \times \left(1\dfrac{7}{8}\right)^2$

$=$

Your skills are strong!

Date / /

Name

Score
/100

1 Calculate.

5 points per question

Don't forget!

According to the order of operations,
- **calculate exponents and numbers in parentheses and brackets first**
- perform multiplication and division before addition and subtraction
- calculate from left to right
---

(1) $1 + 6 - 4 \div 8 =$

(2) $1 + (6 - 4) \div 8 =$

(3) $[(1 + 6) - 4] \div 8 =$

(4) $\frac{1}{2} \div \frac{1}{3} \times 2 + 7 =$

(5) $\frac{1}{2} \div \left(\frac{1}{3} \times 2\right) + 7 =$

(6) $\left(\frac{1}{2} \div \frac{1}{3}\right) \times (2 + 7) =$

(7) $\frac{2}{3} + 1\frac{1}{2} \div \frac{1}{4} \times 1\frac{1}{3} =$

(8) $\frac{2}{3} + 1\frac{1}{2} \div \left(\frac{1}{4} \times 1\frac{1}{3}\right) =$

2 **Calculate.**

(1) $2 \div 6 + \left(\dfrac{1}{2} \right)^{2} + 1 =$

(4) $3 \dfrac{1}{3} + 2^{2} - 9 \div 4 \times 2 =$

(2) $6 \div \left(2 + \dfrac{1}{2} \right)^{2} + 1 =$

(5) $\left(1 \dfrac{1}{3} + 2 \right)^{2} - 9 \div (4 \times 2)$

$=$

(3) $6 \div \left[\left(2 + \dfrac{1}{2} \right)^{2} + 1 \right] =$

(6) $\left[\left(1 \dfrac{1}{3} + 2 \right)^{2} - 9 \right] \div (4 \times 2)$

$=$

Nicely done!

7

Negative Numbers Review

Date / /

Name

Score

/100

1 Calculate.

2 points per question

(1) $3 - 7 =$

(2) $-2 + 8 =$

(3) $6 - (-7) =$

(4) $\dfrac{1}{4} + (-1) =$

(5) $-\dfrac{2}{5} + 7 =$

(6) $-2 - \left(-1\dfrac{1}{3}\right) =$

(7) $-2\dfrac{2}{3} - \dfrac{1}{4} =$

(8) $-3\dfrac{2}{5} - \left(-1\dfrac{1}{4}\right) =$

(9) $1 - 2\dfrac{1}{2} =$

(10) $-1 - 2\dfrac{1}{2} =$

(11) $-1 - \left(-2\dfrac{1}{2}\right) =$

(12) $7\dfrac{1}{3} - 2\dfrac{1}{2} =$

(13) $-7\dfrac{1}{3} + \left(-2\dfrac{1}{2}\right) =$

(14) $-7\dfrac{1}{3} - \left(-2\dfrac{1}{2}\right) =$

2 **Calculate.**

(1) $6 \times (-2) =$

(2) $-4 \times (-8) =$

(3) $-\dfrac{4}{5} \times \dfrac{3}{8} =$

(4) $1\dfrac{4}{5} \times \left(-3\dfrac{1}{3}\right) =$

(5) $18 \div (-3) =$

(6) $(-24) \div (-12) =$

(7) $-\dfrac{1}{4} \div 10 =$

(8) $\dfrac{1}{2} \div \left(-\dfrac{3}{8}\right) =$

(9) $9 \times \left(-1\dfrac{2}{3}\right) \div \left(-3\dfrac{1}{3}\right) =$

(10) $-2 \div \left(-2\dfrac{2}{5}\right) \times \left(-1\dfrac{4}{5}\right) =$

(11) $-3\dfrac{1}{5} \div \dfrac{8}{15} \div 9 \times \left(-1\dfrac{1}{8}\right)$

$=$

(12) $6\dfrac{2}{3} \times \left(-2\dfrac{1}{4}\right) \div \left(-3\dfrac{1}{3}\right) \div \left(-1\dfrac{2}{3}\right)$

$=$

When multiplying or dividing with negative numbers, count the number of negative signs and determine the sign of the answer first. Then calculate.

Positively wonderful work!

9

Negative Numbers Review

Date ___ / ___ / ___

Name

Score

/ 100

1 **Calculate.**

5 points per question

(1) $(-2)^5 =$

(2) $(-3)^3 \times (-2)^2 =$

(3) $6^2 \times (-3)^0 =$

(4) $(-8) \times \left(-\dfrac{1}{4}\right)^3 =$

(5) $-\left(-1\dfrac{1}{3}\right)^4 \times \left(-1\dfrac{1}{8}\right)^2 =$

(6) $(-2)^4 \div (-4)^3 =$

(7) $-2^4 \div (-4)^3 =$

(8) $3^3 \div \left(-\dfrac{1}{2}\right)^2 =$

(9) $\left(-\dfrac{3}{5}\right)^2 \div \left(-\dfrac{9}{10}\right)^3 =$

(10) $\left(-2\dfrac{1}{2}\right)^3 \div \left(1\dfrac{7}{8}\right) =$

Pay close attention to the placement of parentheses.

2 **Calculate.**

(1) $\dfrac{-\dfrac{1}{2}}{\dfrac{3}{4}} =$

(2) $\dfrac{\dfrac{3}{7}}{-\dfrac{6}{11}} =$

(3) $\dfrac{8}{\dfrac{1}{3} - \dfrac{1}{2}} =$

(4) $\dfrac{-\dfrac{1}{4} - \dfrac{1}{8}}{5} =$

(5) $\dfrac{\dfrac{1}{4} - \dfrac{3}{5}}{2\dfrac{1}{4} - 1\dfrac{1}{2}} =$

(6) $(-1)^6 \times (-3)^2 \div (-2)^3$

$=$

(7) $3^4 \div (-2)^5 \times 6^0 \div (-3)^2$

$=$

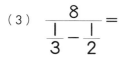

Remember to follow the order of operations.

(8) $(-2) - (-4)^3 \div (-3)^2 + (-5)^2$

$=$

(9) $\left(-\dfrac{2}{5}\right)^2 \times (2 + 6 \times 3) \div (-2)^3$

$=$

(10) $\left(-3\dfrac{2}{3}\right)^0 \div \left(-\dfrac{3}{4} - \dfrac{1}{4} \times 2\right) + (-12 \div 10)^2$

$=$

If you have difficulty with this workbook, please try *Pre-Algebra Workbook II*.

Great job!

6

Values of Algebraic Expressions Review

Date / /

Name

Level ☆

Score

/100

1 Determine the value of each expression when $x = 3$.

5 points per question

(1) $-4x =$

(2) $\dfrac{5}{6}x =$

(3) $x - 9 =$

(4) $2x + 4 =$

(5) $-5x - 6 =$

(6) $\dfrac{1}{2}x - \dfrac{9}{2} =$

(7) $8 \div \dfrac{1}{x} =$

(8) $6 - \dfrac{3}{2x} =$

(9) $8x - \dfrac{5}{2x} =$

(10) $-\dfrac{3}{2}x + \dfrac{x}{4} =$

2   **Determine the value of each expression when** $s = \dfrac{2}{3}$**.**

5 points per question

(1) $5s =$

(2) $2s - \dfrac{1}{4} =$

(3) $6 - \dfrac{1}{4}s =$

(4) $\dfrac{s}{4} =$

(5) $-\dfrac{6}{s} =$

(6) $\dfrac{2-s}{s} =$

(7) $\dfrac{4-s}{3s+2} =$

(8) $\dfrac{s}{2} - 3 - \dfrac{1}{s} =$

(9) $-\dfrac{7}{10}s \div 2s =$

(10) $\dfrac{s + \dfrac{1}{2}}{1 - s} =$

Keep up the good effort!

1 **Determine the value of each expression when $j = -2$.**

5 points per question

(1) $j^5 =$

(2) $-j^3 =$

(3) $3 - j^2 =$

(4) $\dfrac{3}{j^2} =$

(5) $\dfrac{-9-j}{j^2+1} =$

(6) $j^2 - 9 =$

(7) $(j+3)(j-3) =$

(8) $j^4 - j^3 =$

(9) $j^5 - j^2 =$

(10) $j^2(j^3 - 1) =$

Pay close attention to the placement of negative signs.

2 **Determine the value of each expression when** $z = -\dfrac{1}{4}$.

5 points per question

(1) $3\dfrac{1}{5} + z =$

(2) $4z - 6 =$

(3) $\dfrac{z}{6} =$

(4) $\dfrac{10}{z+1} =$

(5) $z^2 + z =$

(6) $z^3 - z =$

(7) $\dfrac{2}{z} + \dfrac{3}{z^2} =$

(8) $\dfrac{z^2+1}{z} - z =$

(9) $2z^2 + 5z + 2 =$

(10) $(2z+1)(z+2) =$

Smart thinking!

Date   /   /

Name

Score

/ 100

---

1   **Determine the value of each expression when** $x = 1$ **and** $y = -2$.   5 points per question

(1) $2x + 3y =$

(2) $y - \dfrac{1}{2}x =$

(3) $-\dfrac{3}{4}x + \dfrac{1}{3}y =$

(4) $-\dfrac{9}{2}x - 2y =$

(5) $-\dfrac{3}{2}y + \dfrac{9}{4}x =$

(6) $4xy + \dfrac{1}{x} =$

(7) $\dfrac{y}{3x} - x =$

(8) $9x^2 - 4y^2 =$

2 **Determine the value of each expression when** $a = -1$, $b = 2$,
and $c = -\dfrac{1}{2}$.

(1) $\dfrac{a}{b} + \dfrac{b}{c} + \dfrac{c}{a} =$

(4) $-(-a)^2 + (-b)^2 + c^3 =$

(2) $ab + bc =$

(5) $(b^2 - c^2) - (a^2 - c^2) =$

(3) $b(a + c) =$

(6) $(b + c)(b - c) - (a + c)(a - c)$

$=$

You are doing valuable work!

Values of Algebraic Expressions Review

Date   /   /

Name

Score

/100

1 Determine the value of each expression when $a = -3$, $b = 2$, $c = -1$, and $d = \dfrac{3}{2}$.

5 points per question

(1) $abcd =$

(2) $\dfrac{a+b}{c-d} =$

(3) $\dfrac{a}{b} - \dfrac{c}{d} =$

(4) $\dfrac{ad - bc}{bd} =$

(5) $b^0 - c + d^2 - a^3 =$

(6) $(c - ad)(a^2 + ab + b^2) =$

(7) $[(a \div b) \div (c \times d)]^3 =$

(8) $a \div [b \div (c \times d)]^2 =$

2 Use the given values of w, x, y, and z to determine the value of the expression $\dfrac{x^2y}{z} - \dfrac{z}{w-x}$.

10 points per question

(1) $w=3$, $x=0$, $y=1$, $z=2$

$\dfrac{x^2y}{z} - \dfrac{z}{w-x} =$

(4) $w=1$, $x=\dfrac{1}{2}$, $y=-\dfrac{1}{3}$, $z=-\dfrac{1}{4}$

$\dfrac{x^2y}{z} - \dfrac{z}{w-x} =$

(2) $w=-4$, $x=-3$, $y=-1$, $z=-2$

$\dfrac{x^2y}{z} - \dfrac{z}{w-x} =$

(5) $w=\dfrac{1}{2}$, $x=\dfrac{2}{3}$, $y=\dfrac{1}{4}$, $z=-\dfrac{3}{2}$

$\dfrac{x^2y}{z} - \dfrac{z}{w-x} =$

(3) $w=2$, $x=-1$, $y=1$, $z=-2$

$\dfrac{x^2y}{z} - \dfrac{z}{w-x} =$

(6) $w=-2\dfrac{1}{3}$, $x=-1\dfrac{1}{3}$, $y=\dfrac{1}{2}$, $z=-2$

$\dfrac{x^2y}{z} - \dfrac{z}{w-x} =$

Bravo!

1 Simplify each expression.

4 points per question

Examples

$3x + 4x = 7x$ $\qquad$ $8a - 2a = 6a$

(1) $2x + 3x = 5\boxed{}$

(2) $6y + 9y =$

(3) $8d + 0 =$

(4) $x + 3x =$

(5) $2z + 5z + z =$

(6) $9b - 4b =$

(7) $10h - 7h =$

(8) $5x - 4x =$

Always write x or $-x$, not $1x$ or $-1x$.

(9) $12y - 5y - 4y =$

(10) $15w - 6w + 13w =$

Don't forget!

A variable is a letter or symbol that represents an unknown amount. Variables can be any symbol or letter, such as x, y, a, $\Diamond$, etc.

2 Simplify each expression.

Example

$4a - a = 3a$

It's okay to leave your answers as improper fractions when there is a variable.

(1) $5x + 9x =$

(2) $7y - 4y =$

(3) $5a - (-4a) = 5a + \boxed{} =$

(4) $-3h + (-10h) =$

(5) $6a - (-a) =$

(6) $5j - 0 =$

(7) $0 - 5j =$

(8) $\dfrac{5}{7}w - \dfrac{3}{7}w =$

(9) $-\dfrac{2}{9}z + \dfrac{7}{9}z =$

(10) $\dfrac{1}{4}f + \left(-\dfrac{3}{4}f\right) =$

(11) $2x + \dfrac{1}{3}x = \dfrac{\boxed{}}{3}x + \dfrac{1}{3}x = \dfrac{\boxed{}}{3}x$

(12) $\dfrac{3}{4}y + 2y =$

(13) $3b - \dfrac{2}{3}b =$

(14) $-\dfrac{1}{4}m - \left(-\dfrac{5}{4}m\right) =$

(15) $7xy - 4xy = \boxed{}xy$

Don't forget to reduce your answers!

Great start on a tough topic!

1 Simplify each expression.

4 points per question

Don't forget!

Combine like terms.   **Example** $3x + 4 + 2x + 7 = 5x + 11$

(1) $2x + 4x + 5 = \boxed{}x + 5$

(2) $10y - 3 + 4y =$

(3) $-6 + \frac{7}{3}h - \frac{1}{3}h =$

(4) $\frac{3}{4}bc + \frac{7}{6}bc + \frac{2}{7} =$

(5) $-\frac{1}{2}ab - \frac{5}{6} - \frac{3}{4}ab =$

(6) $-5x - 3 + 8 - 2x =$

(7) $-2x + 5x - 3 - 8 =$

(8) $2a + 4b + 3b - 6a = \boxed{}a + \boxed{}b$

(9) $-\frac{5}{3}b + 2a + \frac{1}{3}b - a =$

(10) $4w - (-w) - y + 2y =$

(11) $\frac{4}{3}x - 2a + \left(-\frac{1}{6}a\right) - \frac{7}{3}x =$

(12) $\frac{7}{2}gh - \frac{1}{3}fg + \left(-\frac{3}{2}gh\right) - \left(-\frac{8}{3}fg\right)$

$=$

(13) $-\frac{9}{8}c - \frac{3}{5}h + \frac{7}{2}c - \left(-\frac{3}{10}h\right)$

$=$

In your answer, write the variables in alphabetical order.

2 **Simplify each expression.**

(1) $3x^2 + 4x^2 - 8x = \boxed{}x^2 - 8x$

 You can combine like terms.

(2) $5x^2 - 2x^2 - 10x + 3x =$

(3) $-5 + 3y^2 + \dfrac{7}{2} - \dfrac{3}{4}y^2 =$

(4) $a^2 + bc - \dfrac{1}{2}bc - \dfrac{10}{3}a^2 =$

(5) $2s + (-v^3) - 3s - (-7v^3) =$

(6) $-\dfrac{7}{3}w + 4n^2 - (-n^2) + (-w)$

=

(7) $2a + 3b + 4c + 6a + 7b + 9c$

$= \boxed{}a + \boxed{}b + \boxed{}c$

(8) $x - 5y - 3x + 2 - \dfrac{1}{2} + \dfrac{3}{4}y$

=

(9) $-2c^2 + b - \dfrac{2}{3} + 9b - 4c^2 + \dfrac{3}{2}$

=

(10) $2x^2 + 5x + 7x^2 + 6 - 5x - 2 = \boxed{}x^2 + \boxed{}$

(11) $-(-8y^2) + 10 - 5y^2 + \left(-\dfrac{5}{2}\right) + \dfrac{9}{2}$

=

(12) $-5abc - (-2be) + 6be + \dfrac{9}{4}abc - \dfrac{5}{2}be$

$= \boxed{}abc + \boxed{}be$

You make this look simple!

23

12 Simplifying Algebraic Expressions

Level ★

Date   /   /

Name

Score

/100

1 **Simplify each expression.**

5 points per question

Examples

$$(x-2y)+(3x+4y)=x-2y+3x+4y=4x+2y$$
$$(6x-9y)+(-2x-8y)=6x-9y-2x-8y=4x-17y$$

(1) $(3x+4y)+(x-6y)$

=

(2) $(a-b)+(-3a-7b)$

=

(3) $(7g-h)+(-6g+4h)$

=

(4) $(-m+3n)+(-m+3n)$

=

(5) $\left(\dfrac{1}{2}a+2b\right)+\left(-2a+\dfrac{3}{2}b\right)$

=

(6) $\left(-\dfrac{2}{3}k-\dfrac{1}{4}l\right)+\left(-\dfrac{1}{2}k+l\right)$

=

(7) $\left(8x-\dfrac{3}{2}\right)+(-9x+4)$

=

(8) $\left(-\dfrac{1}{3}s-\dfrac{1}{4}t\right)+\left(-\dfrac{1}{2}s-\dfrac{3}{5}t\right)$

=

2 **Simplify each expression.**

10 points per question

(1) $(3a-2b+6c)+(-a-3b+4c)$

=

(2) $(-2x-y+3z)+(5x-4y-2z)$

=

(3) $\left(a+2c-\dfrac{2}{3}d\right)+\left(a-\dfrac{3}{4}c+\dfrac{2}{3}d\right)$

=

(4) $\left(-\dfrac{3}{4}b+\dfrac{5}{2}c^2+\dfrac{1}{3}de\right)+\left(-\dfrac{1}{4}c^2-\dfrac{5}{4}de-2b\right)$

=

It's okay to leave your answers as improper fractions when there is a variable.

(5) $(2a+9b+3c-5)+(3a-4b-6c+7)$

=

(6) $\left(-f+\dfrac{9}{4}g-\dfrac{5}{3}h+e\right)+\left(\dfrac{3}{2}h-\dfrac{1}{2}f-\dfrac{3}{4}e-\dfrac{3}{4}g\right)$

=

Try to concentrate. Some questions are long.

Simplifying Algebraic Expressions

Date / /

Name

Score /100

1 Write either + or − in each of the boxes.

5 points per question

Example $x-(y+z)=x\boxed{-}y\boxed{-}z$

(1) $x-(y-z)=x\boxed{}y\boxed{}z$

(5) $x+(-y-z)=x\boxed{}y\boxed{}z$

(2) $a+(-b+c)=a\boxed{}b\boxed{}c$

(6) $a-(-b-c)=a\boxed{}b\boxed{}c$

(3) $-g-(-h-i)=-g\boxed{}h\boxed{}i$

(7) $g-(h+2i)=g\boxed{}h\boxed{}2i$

(4) $-l-(-m+n)=-l\boxed{}m\boxed{}n$

(8) $l+(-3m-n)=l\boxed{}3m\boxed{}n$

2 Simplify each expression.

6 points per question

Example $5x-(2x+4)=5x-2x-4=3x-4$

Write the correct sign when removing the parentheses.

(1) $9x-(4x-3)=$

(2) $9x-(-4x-3)=$

(3) $9x-(-4x+3)=$

(4) $-5x-(3x-7)=$

(5) $-5x-(-3x+7)=$

(6) $-5x+(-3x-7)=$

(7) $4a+\left(\dfrac{5}{2}-\dfrac{1}{2}a\right)=$

(8) $-4a-\left(\dfrac{5}{2}-\dfrac{1}{2}a\right)=$

(9) $-\dfrac{3}{2}b-(-a+b)=$

(10) $-\dfrac{7}{3}y-\left(-y-\dfrac{3}{4}x\right)=$

Fantastic!

Simplifying Algebraic Expressions

Level

Date   /   /

Name

Score   /100

1 **Simplify each expression.**

5 points per question

Example $(9x+4y)-(2x-6y)=9x+4y-2x+6y=7x+10y$

(1) $(5x-4y)-(3x-6y)$

=

(2) $(-5x+4y)-(-3x+6y)$

=

(3) $(5x+4y)-(-3x+6y)$

=

(4) $(-5x-4y)-(-3x-6y)$

=

(5) $(-5x-4y)-(3x+6y)$

=

(6) $(a-2b)-(3a-b)$

=

(7) $(a+2b)-(3a+b)$

=

(8) $(-a+2b)-(3a-b)$

=

(9) $(-a-2b)-(-3a+b)$

=

(10) $(-a+2b)-(3a+b)$

=

> Remember to write a or $-a$, not $1a$ or $-1a$.

2 **Simplify each expression.**

5 points per question

(1) $(2a-3b)+(4a-9b)$

=

(2) $(2a-3b)-(4a-9b)$

=

(3) $(-2a+3b)-(-4a-9b)$

=

(4) $(-2a-3b)+(-4a+9b)$

=

(5) $(-3x+2y)+(2x+2y)$

=

(6) $\left(-\dfrac{1}{2}f+3g\right)-\left(\dfrac{1}{3}g+\dfrac{1}{4}f\right)$

=

(7) $\left(-\dfrac{1}{3}y-\dfrac{2}{5}z\right)-\left(\dfrac{3}{5}y-\dfrac{1}{4}z\right)$

=

(8) $\left(\dfrac{3}{4}m-\dfrac{7}{3}n\right)-\left(-\dfrac{1}{3}m-\dfrac{3}{4}n\right)$

=

(9) $\left(-\dfrac{5}{2}a+\dfrac{3}{2}b\right)-\left(-\dfrac{7}{5}b+\dfrac{5}{4}a\right)$

=

(10) $\left(-\dfrac{6}{5}x-\dfrac{3}{8}y\right)-\left(-\dfrac{5}{2}y-\dfrac{8}{3}x\right)$

=

Terrific thinking!

Simplifying Algebraic Expressions

Date   /   /

Name

Score

/ 100

1 **Simplify each expression.**

5 points per question

Examples

$$\begin{array}{r} 2x + 5y \\ +)\,4x + 7y \\ \hline 6x + 12y \end{array}$$

The above example is the same as $(2x+5y)+(4x+7y)$

$$\begin{array}{r} 6a + 11b \\ -)\,2a + 8b \\ \hline 4a + 3b \end{array}$$

The above example is the same as $(6a+11b)-(2a+8b)$

(1)
$$\begin{array}{r} 3x + 2y \\ +)\ \ x + 4y \\ \hline 4x + 6y \end{array}$$

(5)
$$\begin{array}{r} 3b + 7c \\ -)\ \ b - 6c \\ \hline \end{array}$$

 This is another way to combine like terms.

(2)
$$\begin{array}{r} 5x + \ y \\ +)\ \ x - 5y \\ \hline \end{array}$$

(6)
$$\begin{array}{r} x - 9y \\ -)\,-5x + 6y \\ \hline \end{array}$$

(3)
$$\begin{array}{r} 2x - 3y \\ +)\,4x - 2y \\ \hline \end{array}$$

(7)
$$\begin{array}{r} -\dfrac{1}{3}s + 3t \\ -)\,-\dfrac{1}{3}s - 2t \\ \hline \boxed{}\,t \end{array}$$

(4)
$$\begin{array}{r} -\dfrac{5}{7}x + 2y \\ +)\ \ \dfrac{1}{7}x - 5y \\ \hline \end{array}$$

(8)
$$\begin{array}{r} \dfrac{7}{4}a - 6b \\ -)\,-\dfrac{1}{4}a - 6b \\ \hline \boxed{}\,a \end{array}$$

2 **Simplify each expression.**

6 points per question

(1)
$$\begin{array}{r} 3a \quad\;\; +c \\ +)\,7a+8b+4c \\ \hline 10a+8b+5c \end{array}$$

(6)
$$\begin{array}{r} 17x+5y \\ -)\;\;4x+2y-8z \\ \hline 13x+3y+\Box z \end{array}$$

Pay close attention to the signs.

(2)
$$\begin{array}{r} 9a-3b \\ +)\,2a+7b-5c \\ \hline \end{array}$$

(7)
$$\begin{array}{r} 5x \quad\;\; -9z \\ -)\,3x-4y \\ \hline \end{array}$$

(3)
$$\begin{array}{r} -4x \quad\;\; +z \\ +)-3x-2y \\ \hline \end{array}$$

(8)
$$\begin{array}{r} -2b+3c \\ -)-a+\;\;b-5c \\ \hline \end{array}$$

(4)
$$\begin{array}{r} -\dfrac{3}{2}x^2-\dfrac{1}{3}x+\dfrac{1}{4} \\ +)\qquad\quad \dfrac{1}{5}x-3 \\ \hline \end{array}$$

(9)
$$\begin{array}{r} \dfrac{1}{2}x^2-\dfrac{1}{3}xy+\dfrac{1}{4}y^2 \\ -)\,\dfrac{1}{3}x^2 \qquad\quad -\dfrac{1}{5}y^2 \\ \hline \end{array}$$

(5)
$$\begin{array}{r} -9x^2+\dfrac{3}{2}xy-\dfrac{1}{3}y^2 \\ +)\;\;9x^2-\dfrac{3}{2}xy+\dfrac{1}{3}y^2 \\ \hline \Box \end{array}$$

(10)
$$\begin{array}{r} \dfrac{7}{12}a^2-2a \\ -)\qquad -\dfrac{2}{3}a-5 \\ \hline \end{array}$$

You're doing a nice job.

31

Simplifying Algebraic Expressions

Date   /   /

Name

---

1 **Simplify each expression by using the distributive property.**

4 points per question

Distributive Property

$$a \times b$$

$$a(b+c) = ab + ac$$

$$a \times c$$

The **distributive property** allows you to simplify the expression by multiplying the terms and removing the parentheses.

(1) $c(d+e) = cd + \boxed{}$

(6) $3x + 2(x+y) = 3x + \boxed{}x + 2y =$

(2) $x(y+z) = \boxed{} + \boxed{}$

(7) $4x - 3(x-y) =$

(3) $a(b-c) = \boxed{} - \boxed{}$

(8) $-2x - (4x - 6y) =$

 Be careful to use the correct sign!

(4) $2(x-y) = 2x - 2y$

(9) $-\dfrac{1}{2}a + 2\left(-a + \dfrac{1}{3}b\right) =$

(5) $-3(-s+t) =$

(10) $\dfrac{3}{4}m - \left(-\dfrac{1}{3}n - \dfrac{1}{2}m\right) =$

---

2 **Simplify each expression.**

Example $2(3x+4y)+5(x-6y)=6x+8y+5x-30y=11x-22y$

(1) $3(2x-4y)+2(3x-5y)$

$\quad =6x-12y+\boxed{}x-\boxed{}y=$

(2) $4(4x-y)+6(-2x+y)$

$\quad =$

(3) $-3(-3x+5y)+(-x+4y)$

$\quad =$

(4) $6\left(\dfrac{1}{2}a-\dfrac{1}{4}b\right)+2\left(-\dfrac{1}{3}a+\dfrac{1}{4}b\right)$

$\quad =$

(5) $\dfrac{1}{2}(-3a^2-4a)+\dfrac{1}{3}(3a-5a^2)$

$\quad =$

(6) $\dfrac{5}{4}(2x^2+4)+\dfrac{5}{2}(3x^2+8)$

$\quad =$

(7) $4(5x+6y)-2(3x+8y)$

$\quad =$

(8) $2(-3x-3y)-(-5x+7y)$

$\quad =$

(9) $-(-g-h)-4(2g-3h)$

$\quad =$

(10) $3\left(-\dfrac{1}{2}a+b\right)-6\left(a-\dfrac{3}{4}b\right)$

$\quad =$

(11) $\dfrac{1}{2}\left(-5c+\dfrac{1}{3}d\right)-\dfrac{1}{3}\left(\dfrac{1}{4}d-\dfrac{3}{5}c\right)$

$\quad =$

(12) $2(2x+3y-z)-\dfrac{1}{2}\left(-x-y+\dfrac{1}{3}z\right)$

$\quad =$

Simply great work!

1 **Simplify each expression.**

5 points per question

Examples

$$\frac{2x+3}{7}+\frac{4x+10}{7}=\frac{(2x+3)+(4x+10)}{7}=\frac{2x+3+4x+10}{7}=\frac{6x+13}{7}$$

$$\frac{8x+11y}{9}-\frac{4x-2y}{9}=\frac{(8x+11y)-(4x-2y)}{9}=\frac{8x+11y-4x+2y}{9}=\frac{4x+13y}{9}$$

(1) $\dfrac{6x+9}{4}+\dfrac{x-6}{4}=\dfrac{(\boxed{})+(\boxed{})}{4}$

$=$

(2) $\dfrac{3x-2}{5}+\dfrac{-6x+9}{5}=$

(3) $\dfrac{5x+11}{3}-\dfrac{3x+4}{3}$

$=\dfrac{(\boxed{})-(\boxed{})}{3}=$

(4) $\dfrac{-3x+5}{6}-\dfrac{-4x-9}{6}=$

(5) $-\dfrac{4x+7}{2}-\dfrac{3x-2}{2}$

$=\dfrac{-(\boxed{})-(\boxed{})}{2}=$

(6) $-\dfrac{11x-8}{7}+\dfrac{-3x+4}{7}=$

(7) $\dfrac{3x+5}{4}-\dfrac{9x}{4}=$

(8) $-\dfrac{-x+3}{8}-\dfrac{5}{8}=$

2 **Simplify each expression.**

6 points per question

Example

$$\frac{x-4}{3}+\frac{x-1}{6}=\frac{2(x-4)+(x-1)}{6}=\frac{3x-9}{6}=\frac{x-3}{2}$$

In this example, the last step is to reduce by using the common factor 3.

(1) $\dfrac{2x+3}{3}+\dfrac{4x+5}{4}$

$=\dfrac{\boxed{}(2x+3)+\boxed{}(4x+5)}{12}$

$=$

(6) $\dfrac{x-3}{2}+\dfrac{5x+7}{4}=$

(2) $\dfrac{4x-5}{2}+\dfrac{3x+2}{3}=$

(7) $\dfrac{-7x+4}{6}-\dfrac{x+1}{2}=$

(3) $\dfrac{-6x+2}{5}+\dfrac{4x-3}{2}=$

(8) $-\dfrac{2x-1}{3}+\dfrac{7x-8}{9}=$

(4) $\dfrac{x+3}{4}+\dfrac{2x+5}{6}=$

$=\dfrac{\boxed{}(x+3)+\boxed{}(2x+5)}{12}=$

(9) $\dfrac{x+5}{2}+\dfrac{7x-11}{6}=$

Reduce your answer!

(5) $\dfrac{2x-5}{6}-\dfrac{-3x-4}{9}=$

(10) $\dfrac{2x-9}{15}-\dfrac{x-3}{3}=$

You can use the least common multiple (LCM) of the denominators.

Keep doing your best!

Solving Equations

Level ★★

Score

/ 100

Date / /

Name

1 **Solve each equation.**

2 points per question

Examples

$x - 3 = 5$
$x = 5 + 3$
$x = 8$

$2 + x = 6$
$x = 6 - 2$
$x = 4$

- To solve equations, think "x = a number," and find that number. For questions such as the examples above, add or subtract from both sides of the equation to find the value of x. To find the value of x in the first example, we add 3 to both sides of the equation: $x - 3 \underline{+ 3} = 5 \underline{+ 3}$.
- You can check if your answer is correct by substituting your value for x into the original equation. For example, if you substitute $x = 8$ into the original equation, the value is $8 - 3 = 5$. Therefore, we know that the answer $x = 8$ is correct.

(1) $x - 2 = 7$

$x = 7 + \boxed{}$

$x =$

(2) $-5 + x = -8$

$x = -8 + \boxed{}$

$x = \boxed{}$

(3) $4 + x = 1$

(4) $-10 + x = -2$

(5) $x + 9 = 2$

(6) $x + 5 = -1$

(7) $x - 6 = 9$

(8) $3 + x = 6$

(9) $-10 + x = -4$

(10) $-1 + x = -1$

2 **Solve each equation.**

5 points per question

(1) $x + 3 = 7$

(2) $x + 6 = 2$

(3) $5 + x = 0$

(4) $-3 + x = -3$

(5) $x - 0.6 = 2$

(6) $1 + x = 2.5$

(7) $x - 2.9 = 1.2$

(8) $x + 1.9 = -1.6$

(9) $-\dfrac{2}{3} + x = -2$

(10) $x - 3 = -\dfrac{3}{4}$

(11) $5 + x = -\dfrac{3}{2}$

(12) $x - \dfrac{1}{2} = \dfrac{1}{3}$

(13) $\dfrac{1}{4} + x = \dfrac{1}{5}$

(14) $x - \dfrac{3}{4} = \dfrac{5}{2}$

(15) $-\dfrac{9}{4} + x = -\dfrac{3}{2}$

(16) $x + \dfrac{5}{6} = \dfrac{4}{3}$

Wow! You solved them!

Solving Equations

Date / /

Name

Score

/100

1 Solve each equation.

5 points per question

Examples

$$2x = 8$$

$$x = 8 \times \frac{1}{2}$$

$$x = 4$$

$$\frac{1}{3}x = -5$$

$$x = -5 \times 3$$

$$x = -15$$

For questions such as the examples above, multiply both sides of the equation to find the value of x. For example, to find the value of x in the first equation, multiply both sides by the **reciprocal** of 2:

$$\frac{1}{2} \times 2x = 8 \times \frac{1}{2}$$

A reciprocal is a fraction flipped upside down.

(1) $5x = 30$

$$x = 30 \times \frac{1}{\boxed{}}$$

$$x =$$

(2) $2x = 14$

(3) $-x = 6$

$$x = 6 \times \left(-\boxed{}\right)$$

$$x =$$

(4) $-3x = 18$

$$x = 18 \times \left(-\boxed{}\right)$$

$$x =$$

(5) $-2x = 7$

(6) $\frac{1}{2}x = 6$

(7) $-\frac{1}{5}x = 2$

(8) $\frac{1}{3}x = \frac{1}{4}$

(9) $-\frac{1}{6}x = 3$

Think of the question as "$x = a$ number."

(10) $\frac{1}{6}x = \frac{3}{4}$

2 Solve each equation.

5 points per question

Example

$$-\frac{2}{3}x = 8$$

$$x = 8 \times \left(-\frac{3}{2}\right)$$

$$x = -12$$

(1) $-\frac{3}{4}x = 6$

(2) $\frac{3}{5}x = 9$

(3) $-\frac{5}{6}x = -10$

(4) $\frac{2}{3}x = 5$

(5) $-\frac{6}{7}x = 8$

(6) $\frac{2}{3}x = -\frac{1}{8}$

(7) $-\frac{2}{9}x = \frac{1}{12}$

(8) $\frac{2}{3}x = \frac{4}{5}$

(9) $-\frac{8}{11}x = -\frac{6}{7}$

(10) $-\frac{7}{2}x = \frac{5}{4}$

You're getting the hang of this!

20 Solving Equations

Level ★★

Date   /   /

Name

Score
/100

1 Solve each equation.

6 points per question

Examples

$2x - 5 = 7$

$2x = 7 + 5$

$2x = 12$

$x = 12 \times \dfrac{1}{2}$

$x = 6$

$\dfrac{2}{3}x + 7 = -1$

$\dfrac{2}{3}x = -1 - 7$

$\dfrac{2}{3}x = -8$

$x = -8 \times \dfrac{3}{2}$

$x = -12$

(1) $3x - 2 = 10$

$3x = 10 + \boxed{}$

$3x = \boxed{}$

$x = \boxed{} \times \dfrac{1}{\boxed{}}$

$x =$

(2) $-5x - 8 = 7$

(3) $5 - 8x = -1$

(4) $6x - 4 = 10$

(5) $-7x + 1 = 8$

(6) $\dfrac{3}{5}x - 4 = 2$

$\dfrac{3}{5}x = 2 + \boxed{}$

$\dfrac{3}{5}x = \boxed{}$

$x = \boxed{} \times \dfrac{5}{\boxed{}}$

$x =$

② **Solve each equation.**

Examples

$$6x = 4x - 10$$
$$6x - 4x = -10$$
$$2x = -10$$
$$x = -10 \times \frac{1}{2}$$
$$x = -5$$

$$10x + 6 = 8x$$
$$10x - 8x + 6 = 0$$
$$2x + 6 = 0$$
$$2x = -6$$
$$x = -6 \times \frac{1}{2}$$
$$x = -3$$

(1) $5x = 2x + 6$

(2) $7x = x - 2$

(3) $4x = 10x + 5$

(4) $3x = -2x + 4$

(5) $6x - 4 = 8x$

(6) $-3x - 1 = 3x$

(7) $\frac{1}{3}x - 4 = \frac{1}{2}x$

(8) $-\frac{1}{4}x - 3 = -\frac{3}{2}x$

Don't give up!

1 Solve each equation.

10 points per question

Example

$$5x + 7 = 3x + 13$$

$$5x - 3x + 7 = 13$$

$$2x + 7 = 13$$

$$2x = 13 - 7$$

$$2x = 6$$

$$x = 6 \times \frac{1}{2}$$

$$x = 3$$

In the example:
- "$3x$" on the right of the equal sign has been transposed to "$-3x$" on the left
- "$+7$" on the left of the equal sign has been transposed to "-7" on the right

Transposition (or **Inverse Operations**) is when we move a term to the other side of the equation and change the sign.

(1) $2x - 7 = -6x + 9$

$$2x + 6x - \boxed{} = 9$$

$$\boxed{}x - \boxed{} = 9$$

(3) $5x - 1 = 7x + 11$

(2) $x - 5 = 2x - 1$

(4) $-2x - 4 = 2x + 8$

2 **Solve each equation.**

10 points per question

(1) $3x + 8 = -2x + 7$

(4) $\dfrac{1}{2}x + 4 = \dfrac{1}{3}x + 1$

(2) $2x - 8 = -11x - 10$

(5) $-\dfrac{2}{5}x + \dfrac{1}{3} = -\dfrac{1}{2}x + \dfrac{1}{4}$

(3) $6x - 6 = 3x + 1$

(6) $\dfrac{5}{3}x - 2 = \dfrac{1}{2}x - \dfrac{2}{3}$

You got to the answer step-by-step!

Solving Equations

Level ★★

Date / /

Name

Score
/ 100

1 **Solve each equation.**

6 points per question

Example

$$7x - 6 = 5x + 10$$
$$7x - 5x = 10 + 6$$
$$2x = 16$$
$$x = 8$$

(1) $9x - 8 = 6x + 4$

$9x \boxed{} 6x = 4 \boxed{} 8$

(4) $-2x + 11 = 2x + 3$

(2) $7x + 2 = 2x - 13$

$7x \boxed{} 2x = -13 \boxed{} 2$

(5) $-3x - 2 = 5x + 6$

(3) $7x - 2 = -5x - 14$

(6) $4x + 2 = -2x - 10$

2 **Solve each equation.**

8 points per question

(1) $-9x + 2 = -6x - 7$

(5) $6x + 2 = 4x + 3$

(2) $8x - 6 = 9x - 2$

(6) $-2x - 5 = x - 7$

(3) $5x + 3 = 3x - 6$

(7) $2x - 2 = \dfrac{1}{2}x + 6$

(4) $5x - 7 = -3x + 9$

(8) $x + \dfrac{2}{3} = -\dfrac{1}{3}x - \dfrac{1}{6}$

Your work is getting better and better!

23

Solving Equations

Level

Date / /

Name

Score

/100

1 Solve each equation, and check your answer.

10 points per question

Example

$$5x - 1 = 3x - 7$$
$$5x - 3x = -7 + 1$$
$$2x = -6$$
$$x = -3$$

Check your answer!

Left side $= 5 \times (-3) - 1 = -15 - 1 = -16$

Right side $= 3 \times (-3) - 7 = -9 - 7 = -16$

If the value on the left side of the equal sign matches the value on the right, then your answer is correct!

(1) $2x - 6 = -3x + 4$

$2x + \boxed{} = 4 + \boxed{}$

Check!

Left side $= 2 \times \boxed{} - 6 =$

Right side $= -3 \times \boxed{} + 4 =$

(2) $4x + 7 = 9 + 5x$

Check!

Left side $=$

Right side $=$

2 **Solve each equation, and check your answer.**

20 points per question

(1) $3x - 3 = -3x - 15$

Check!

Left side =

Right side =

(2) $1 + \dfrac{2}{3}x = x - 1$

Check!

Left side =

Right side =

(3) $\dfrac{3}{4}x - 1 = -\dfrac{3}{2}x + 2$

Check!

Left side =

Right side =

(4) $-\dfrac{1}{2}x + \dfrac{1}{5} = -3x + \dfrac{1}{2}$

Check!

Left side =

Right side =

By checking, you're sure to get every answer correct!

Solving Equations

Date / /

Name

Score

/100

---

1 Solve each equation.

6 points per question

Example

$$3(4x+6)=2(5x+7)$$

To simplify, remove the parentheses.

$$12x+18=10x+14$$
$$12x-10x=14-18$$
$$2x=-4$$
$$x=-2$$

(1) $2(x-5)=-4$

$$2x-\boxed{}=-4$$

(4) $-2(x+1)=-6(x-1)$

(2) $3x=4(x+2)$

(5) $-(x-5)=3(7-x)$

(3) $-(2-4x)=-(x-3)$

(6) $4(-6+x)=-3(2x-7)$

2 **Solve each equation.**

8 points per question

(1) $4+2(x+4)=-x$

$4+\boxed{}x+\boxed{}=-x$

(5) $-(x+5)=2(2x-3)-7$

(2) $2-3(2x+3)=x$

(6) $6\left(\dfrac{1}{2}x+1\right)-7=-3\left(-\dfrac{1}{2}x-1\right)$

(3) $-x-(3x-2)=14$

(7) $2\left(-\dfrac{3}{4}x-\dfrac{1}{2}\right)-2x=-\left(-\dfrac{1}{3}+x\right)$

(4) $4(-x-2)-3x=-x$

(8) $-\dfrac{2}{3}(3+2x)+4=-(1+x)-\dfrac{1}{2}x$

You're a champ!

1 Solve each equation.

10 points per question

Example

$$\frac{1}{3}x + \frac{1}{6} = \frac{1}{2}x + \frac{5}{6}$$

$$\left(\frac{1}{3}x + \frac{1}{6}\right) \times 6 = \left(\frac{1}{2}x + \frac{5}{6}\right) \times 6$$

$$2x + 1 = 3x + 5$$

$$2x - 3x = 5 - 1$$

$$-x = 4$$

$$x = -4$$

To remove the denominator and solve for x, multiply each side by the LCM of the denominators—in this example, it is 6.

(1) $\frac{1}{4}x + \frac{1}{2} = \frac{1}{6}x + \frac{1}{3}$

Multiply each side by $\boxed{}$:

$$\left(\frac{1}{4}x + \frac{1}{2}\right) \times \boxed{} = \left(\frac{1}{6}x + \frac{1}{3}\right) \times \boxed{}$$

(3) $5 - \frac{1}{4}x = 7 - \frac{1}{6}x$

(2) $\frac{7}{10}x + \frac{2}{5} = \frac{7}{20} + \frac{3}{4}x$

(4) $-\frac{5}{6}x - 2 = \frac{1}{2} - \frac{3}{8}x$

2 **Solve each equation.**

Example

$$\frac{x+1}{3} = \frac{2x-7}{5}$$

$$\frac{x+1}{3} \times 15 = \frac{2x-7}{5} \times 15$$

$$5(x+1) = 3(2x-7)$$

$$5x+5 = 6x-21$$

$$-x = -26$$

$$x = 26$$

(1) $\dfrac{x-2}{4} = \dfrac{3x-7}{10}$

$\dfrac{x-2}{4} \times \boxed{} = \dfrac{3x-7}{10} \times \boxed{}$

(4) $\dfrac{x}{3} - \dfrac{3-2x}{6} = \dfrac{5}{2}x$

You can always check your answer by plugging the value of x into the original equation.

(2) $\dfrac{-6x-2}{5} = \dfrac{4-3x}{2}$

(5) $-\dfrac{6x-7}{5} + 3 = \dfrac{4-3x}{2}$

(3) $\dfrac{-2x-3}{4} = \dfrac{-6-5x}{6}$

(6) $\dfrac{-3-5x}{4} - 2 = \dfrac{2x-1}{6} - 1$

Extraordinary job!

26

Word Problems with Equations

Level ★★

Date / /

Name

Score

/100

1 **Write each sentence as an equation, and then solve.** 6 points per question

(1) x added to 3 is 7.

$3 + x = \boxed{}$

(2) Adding $\frac{1}{2}$ and x is -3.

(3) Subtracting 2 from x equals 10.

$\boxed{} - 2 = 10$

(4) Subtracting $\frac{1}{2}$ from x equals $\frac{1}{3}$.

(5) Subtracting x from 5 equal 4.

$5 - x = \boxed{}$

(6) x equals 6 times -2.

$x = 6 \times (\boxed{})$

(7) $\frac{3}{4}$ times x equals 6.

(8) x divided by 5 equals 4.

$\frac{x}{\boxed{}} = 4$

(9) x divided by 3 equals $\frac{2}{5}$.

(10) x divided by 12 is $\frac{2}{3}$.

(2) **Write each word problem as an equation, and then solve.** 10 points per question

(1) Patty has a bag of yo-yos that weighs 76 ounces. If each yo-yo weighs 4 ounces, how many yo-yos are in Patty's bag?

If there are x yo-yos,

$4 \times x = \boxed{}$

⟨Ans.⟩ _____ yo-yos

(2) Nathalie purchases a crate of potatoes that weighs 192 ounces. If there are 24 potatoes in the crate, how much does each potato weigh?

If each potato weighs x ounces,

⟨Ans.⟩ _____

(3) Dr. Mary orders a box of apples and oranges that weighs 39 ounces. Each apple weighs 5 ounces and each orange weighs 4 ounces. If there are 3 apples in the box, how many oranges are there?

If there are x oranges,

$5 \times 3 + \boxed{} \times x = 39$

Don't forget to write the units!

⟨Ans.⟩ _____

(4) Eriko buys a bag of peanuts and walnuts that weighs 315 grams. There are 25 peanuts and 40 walnuts in the bag. If each peanut weighs 3 grams, how much does each walnut weigh?

If each walnut weighs x grams,

⟨Ans.⟩ _____

You're a whiz with word problems!

Word Problems with Equations

Level ★★

Date　　/　　/

Name

Score
/100

1　Write each word problem as an equation.　　10 points per question

(1) There are a total of 12 pencils and pens in Hideyo's backpack. If there are x pencils, how many pens are there?

〈Ans.〉 $\left(\boxed{}-x\right)$ pens

(2) There are a total of 30 crayons and pens on Trent's desk. If there are x pens, how many crayons are there?

〈Ans.〉

(3) In Christy's pet shop, there are 2 more dogs than cats. If there are x dogs, how many cats are there?

〈Ans.〉

(4) There are 25 students in Evan's class. If there are x boys, how many girls are there?

〈Ans.〉

(5) In Kevin's class, there are 5 more girls than boys. If there are x boys, how many girls are there?

〈Ans.〉

(6) In Maurice's class, there are also 5 more girls than boys. If there are x girls, how many boys are there?

〈Ans.〉

2 **Write the word problem as two possible equations, and then solve.** 20 points for completion

There are 24 students in Steve's swimming class. There are 6 more girls than boys in the pool. How many boys are in the pool? How many girls are in the pool?

Method #1:

If there are x boys in the pool,

then there are $\left(x + \boxed{} \right)$ girls in the pool.

$x + \left(x + \boxed{} \right) = 24$

〈Ans.〉 boys
_____ girls

Method #2:

If there are x girls in the pool,

then there are $\left(x - \boxed{} \right)$ boys in the pool.

$x + \left(\boxed{} \right) = 24$

〈Ans.〉
_____

3 **Write the word problem as two possible equations, and then solve.** 20 points for completion

Debby bought 31 apples and bananas. There are 9 more apples than bananas. How many apples did she buy? How many bananas did she buy?

〈Ans.〉
_____

〈Ans.〉
_____

I'm impressed!

55

1 Write each word problem as an equation, and then solve.

20 points per question

(1) Jeanine has 22 markers, and Mary has 8 markers. Jeanine gives Mary x markers, so they will each have the same number of markers. How many markers did Jeanine give Mary?

Jeanine gives Mary x markers,

therefore Jeanine has a new amount of $\left(22-\boxed{}\right)$ markers,

and Mary has a new amount of $\left(\boxed{}+x\right)$ markers.

Because they have the same number of markers,

$$22-\boxed{}=\boxed{}+x$$

⟨Ans.⟩ _____

(2) John has 20 coins, and Laura has 6 coins. John gives Laura x coins, so John will only have 4 more coins than Laura. How many coins did John give Laura?

John gives Laura x coins,

therefore John has a new amount of $\left(20-\boxed{}\right)$ coins,

and Laura has a new amount of $\left(6+x\right)$ coins.

Because John has 4 more coins than Laura,

$$20-\boxed{}=\left(6+x\right)+\boxed{}$$

⟨Ans.⟩ _____

2 **Write each word problem as an equation, and then solve.**

(1) Masa is 12 years old. In how many years will Masa be 4 times his current age?

If x is the number of years that pass,

then Masa's new age will be $\left(12+\boxed{}\right)$ years old.

Because Masa will be 4 times his current age,

$$\left(12+\boxed{}\right)=\boxed{}\times 12$$

⟨**Ans.**⟩ _____

(2) In 10 years, Mike will be 3 times his current age. What is Mike's current age?

If x is his current age,

then

⟨**Ans.**⟩ _____

(3) Justin is 10 years old, and Alan is 26 years old. In how many years will Alan be twice as old as Justin?

If x is the number of years that pass,

then Justin's new age will be $\left(10+\boxed{}\right)$ years old,

and Alan's new age will be $\left(26+\boxed{}\right)$ years old.

Because Alan will be twice as old as Justin,

$$\boxed{}\times\left(10+\boxed{}\right)=26+\boxed{}$$

⟨**Ans.**⟩ _____

These are tricky! Nice job!

Solving Equations

Date / /

Name

Score /100

1 Solve each equation for x.

5 points per question

Examples

$x + 3 = b$

$x = b - 3$

Transpose 3 to the right of the equal sign.

$a = b - x$

$x = b - a$

Transpose $-x$ to the left of the equal sign and a to the right.

(1) $x + 6 = b$

$x = b - \boxed{}$

(2) $x + 1 = c$

(3) $x - 8 = a$

$x = a + \boxed{}$

(4) $x - 4 = d$

(5) $x - 11 = -g$

(6) $-a = -b - x$

(7) $c = -d - x$

(8) $-1 = b - x$

$x = b + \boxed{}$

(9) $2 = -y - x$

(10) $-a = -3 - x$

> In your answer, write the variables in alphabetical order.

2 **Solve each equation for** x.

5 points per question

Examples

$$2x = a$$

$$x = \frac{a}{2}$$

$$4x = a + b$$

$$x = \frac{a+b}{4}$$

(1) $ax = c$

(6) $bx = -c + d$

(2) $-3x = b$

$$x = \frac{b}{\boxed{}}$$

$$x = -\frac{b}{\boxed{}}$$

(3) $-fx = -g$

(7) $cx + 3 = d$

$$cx = d - \boxed{}$$

(8) $ax - 5 = b$

(4) $s = 4x$

$$4x = \boxed{}$$

$$x =$$

(9) $s - 2t = rx$

$$rx = \boxed{}$$

(5) $7 = ax$

(10) $-y + 2z = wx$

Hats off to you!

59

30

Solving Equations

Date / /

Name

Level

Score

/ 100

1 **Solve each equation for** x.

10 points per question

Example

$$\frac{x}{a} + \frac{b}{c} = d$$

$$ac\left(\frac{x}{a} + \frac{b}{c}\right) = acd$$

$$cx + ab = acd$$

$$cx = acd - ab$$

$$x = \frac{acd - ab}{c}$$

To remove the denominator and solve for x, multiply each side by the LCM of the denominators—in this example, it is ac.

(1) $\dfrac{x}{a} = b - \dfrac{c}{d}$

Multiply each side by the LCM of the denominators, which is ☐.

(4) $\dfrac{x}{c} = \dfrac{d}{b}$

Multiply each side by the LCM of the denominators, which is ☐.

(2) $\dfrac{x}{6} = -2 - \dfrac{3}{4}x$

(5) $\dfrac{a}{b}x = \dfrac{c}{d}$

(3) $2x + \dfrac{5}{3} = -\dfrac{1}{6}x$

(6) $\dfrac{3f}{g} = -\dfrac{4x}{5h}$

2 **Solve each equation for x.**

5 points per question

(1) $x - a = b$

(5) $\dfrac{1}{3}(x - a) = b$

(2) $2(x - a) = b$

$x - a = \dfrac{b}{\boxed{}}$

(6) $\dfrac{1}{5}(x + 3) = -d$

(3) $y(x + 4) = a$

(7) $\dfrac{1}{a}(3 - 4x) = 5$

(4) $a(x + b) = c + d$

(8) $\dfrac{a(b - d)}{x} = -M$

You go the extra mile, and it shows!

31 Simultaneous Linear Equations

Level ★★★

Score

Date / /

Name

/100

1 **Solve each equation for both variables.**

15 points per question

Example

$$\begin{cases} 4x+5y=7 & \cdots ① \\ x+5y=13 & \cdots ② \end{cases}$$ Step 1: Number each equation.

$$① - ② : \quad \begin{array}{r} 4x+5y=7 \quad \cdots ① \\ -)\ x+5y=13 \quad \cdots ② \\ \hline 3x \qquad =-6 \\ x=-2 \end{array}$$

Step 2: Remove a variable by subtracting one equation from the other.

Substitute this into ② :

$$-2+5y=13$$
$$5y=15$$
$$y=3$$

Step 3: Substitute the value of the variable into either original equation (usually the simpler one) to solve for the other variable.

⟨Ans.⟩ $(x, y)=(-2,\ 3)$ Step 4: Write your answer.

Two linear equations with the same variables are called **simultaneous linear equations**. You can solve for the variables by subtracting one equation from the other, a method known as the **subtraction method**.

(1) $$\begin{cases} x+3y=-13 & \cdots ① \\ 7x+3y=-19 & \cdots ② \end{cases}$$

$$① - ② : -6x = \boxed{}$$
$$x = \boxed{}$$

Substitute this into ① :

$$\boxed{}+3y=-13$$
$$3y=\boxed{}$$
$$y=\boxed{}$$

⟨Ans.⟩ $(x, y)=(\boxed{},\ \boxed{})$

(2) $$\begin{cases} 2x-5y=-17 & \cdots ① \\ 2x+3y=23 & \cdots ② \end{cases}$$

$$① - \boxed{②} :$$

Substitute this into ② :

⟨Ans.⟩

© Kumon Publishing Co., Ltd.

2 **Solve each equation for both variables.**

14 points per question

(1) $\begin{cases} 2x + 3y = 4 \\ 7x + 3y = -1 \end{cases}$

(4) $\begin{cases} -7x - 2y = -38 \\ -7x + 8y = 12 \end{cases}$

⟨Ans.⟩ _____

⟨Ans.⟩ _____

(2) $\begin{cases} 4x + 3y = -15 \\ 4x - y = -27 \end{cases}$

(5) $\begin{cases} -5x - 4y = 17 \\ 3x - 4y = -23 \end{cases}$

⟨Ans.⟩ _____

⟨Ans.⟩ _____

(3) $\begin{cases} 5x - 6y = 4 \\ 3x - 6y = 0 \end{cases}$

Don't forget to number the equations!

This is complex, so concentrate!

⟨Ans.⟩ _____

Simultaneous Linear Equations ★★★

Level

Date / /

Name

Score

/100

1 Solve each equation for both variables.

14 points per question

Example

$$\begin{cases} 5x + 4y = 26 & \cdots ① \\ 3x - 4y = 22 & \cdots ② \end{cases}$$

Step 1: Number each equation.

$$① + ② : \quad 5x + 4y = 26 \quad \cdots ①$$
$$+ \,) \, 3x - 4y = 22 \quad \cdots ②$$
$$\overline{\qquad 8x \qquad = 48}$$
$$x = 6$$

Step 2: Remove a variable by adding one equation to the other.

Substitute this into ① :

$$30 + 4y = 26$$
$$4y = -4$$
$$y = -1$$

Step 3: Substitute the value of the variable into either original equation (usually the simpler one) to solve for the other variable.

⟨Ans.⟩ $(x, y) = (6, -1)$

Step 4: Write your answer.

You can solve for the variables by adding the two equations together, a method known as the **addition method**.

(1) $\begin{cases} 5x - 3y = 13 \\ -x + 3y = -5 \end{cases}$

(2) $\begin{cases} -4x + y = 17 \\ 4x - 8y = 4 \end{cases}$

⟨Ans.⟩ _____

⟨Ans.⟩ _____

2 Solve each equation by adding or subtracting the equations.

18 points per question

(1) $\begin{cases} 5x + 2y = -22 & \cdots ① \\ 3x - 2y = -10 & \cdots ② \end{cases}$

① $\square$ ② :

(3) $\begin{cases} \dfrac{1}{2}x + 4y = 15 \\ \dfrac{1}{2}x - 5y = -12 \end{cases}$

⟨Ans.⟩ _____

⟨Ans.⟩ _____

(2) $\begin{cases} 6x + 2y = -5 \\ 6x - y = 7 \end{cases}$

(4) $\begin{cases} 2x - \dfrac{1}{3}y = 1 \\ \dfrac{1}{2}x + \dfrac{1}{3}y = \dfrac{3}{4} \end{cases}$

⟨Ans.⟩ _____

⟨Ans.⟩ _____

Variables can be fractions, too.

You can solve in two ways now! Wow!

33 Simultaneous Linear Equations ★★★

Level

Score

/100

Date / /

Name

1 Solve each equation.

20 points per question

Example

$$\begin{cases} -3x+y=9 & \cdots① \\ -5x-2y=26 & \cdots② \end{cases}$$

①×2: $\quad -6x+2y=18 \quad \cdots③$

②+③: $\quad -5x-2y=26 \quad \cdots②$

$\quad\quad +)-6x+2y=18 \quad \cdots③$

$\quad\quad -11x \quad\quad =44$

$\quad\quad\quad x=-4$

Substitute this into ① :

$\quad 12+y=9$

$\quad\quad y=-3$

⟨**Ans.**⟩ $(x, y)=(-4, -3)$

Step 1: Number each equation.

Step 2: Multiply one equation by a constant and number the new equation. In this example, we multiplied equation ① by 2 and called it equation ③.

Step 3: Remove a variable by using the addition or subtraction method.

Step 4: Substitute the value of the variable into either original equation (usually the simpler one) to solve for the other variable.

Step 5: Write your answer.

(1) $\begin{cases} 2x+y=11 & \cdots① \\ 5x-2y=5 & \cdots② \end{cases}$

①×2 : $4x+\boxed{}y=\boxed{} \cdots③$

②+③ : $\quad 9x=\boxed{}$

$\quad\quad x=\boxed{}$

Substitute this into ① :

$2\times\boxed{}+y=11$

$\quad y=\boxed{}$

⟨**Ans.**⟩ $(x, y)=(\boxed{}, \boxed{})$

(2) $\begin{cases} 8x-5y=21 & \cdots① \\ 2x+9y=-5 & \cdots② \end{cases}$

②×4 :

⟨**Ans.**⟩ _____

2 **Solve each equation, and check your answer.**

30 points per question

┌─ **Don't forget!** ───┐
│ You can always check your answer by substituting your answer into the original equations and checking if │
│ both equations are satisfied. │
└──┘

(1) $\begin{cases} 2x + 3y = 12 & \cdots ① \\ 5x - y = 13 & \cdots ② \end{cases}$

<u>Check!</u>

① : $2 \times \boxed{} + 3 \times \boxed{} = \boxed{} + \boxed{} = \boxed{}$

The left side of ① matches the right side!

② : $5 \times \boxed{} - \boxed{} = \boxed{} - \boxed{} = \boxed{}$

The left side of ② matches the right side!

Therefore, your answer is correct!

⟨**Ans.**⟩ _____

(2) $\begin{cases} 6x + 5y = -9 & \cdots ① \\ -2x - 3y = 7 & \cdots ② \end{cases}$

<u>Check!</u>

⟨**Ans.**⟩ _____

Double checking your work is smart!

© Kumon Publishing Co., Ltd.   67

34

Simultaneous Linear Equations

Level ★★★

Date / /

Name

Score

/100

1 Solve each equation.

20 points per question

Example

$$\begin{cases} 5x+2y=16 & \cdots① \\ 4x+3y=17 & \cdots② \end{cases}$$

①×3: $15x+6y=48$ ···③

②×2: $8x+6y=34$ ···④

③−④: $15x+6y=48$ ···③

$$\underline{-)\ \ 8x+6y=34 \ \cdots④}$$

$$7x\ \ \ \ \ \ =14$$

$$x=2$$

Substitute this into ② :

$$8+3y=17$$

$$3y=9$$

$$y=3$$

⟨Ans.⟩ $(x, y)=(2, 3)$

Step 1: Number each equation.

Step 2: Multiply one equation by a constant and number the new equation.

Step 3: Multiply the other equation by a constant and number the new equation.

Step 4: Remove a variable by using the addition or subtraction method.

Step 5: Substitute the value of the variable into either original equation (usually the simpler one) to solve for the other variable.

Step 6: Write your answer.

(1) $\begin{cases} 5x+3y=-7 & \cdots① \\ 6x+4y=-8 & \cdots② \end{cases}$

①×4: $\begin{cases} 20x+12y=\boxed{} & \cdots③ \\ 18x+12y=\boxed{} & \cdots④ \end{cases}$

②×$\boxed{}$:

③−④: $2x=\boxed{}$

$x=\boxed{}$

Substitute this into ① :

$$5\times(\boxed{})+3y=-7$$

$$3y=\boxed{}$$

$$y=\boxed{}$$

⟨Ans.⟩ $(x, y)=(\boxed{}, \boxed{})$

(2) $\begin{cases} -2x+3y=-2 & \cdots① \\ 5x-7y=6 & \ \ \ \ \cdots② \end{cases}$

①×5: $\Big\{$

②×$\boxed{}$: $\Big\{$

⟨Ans.⟩ _____

2 **Solve each equation, and check your answer.**

20 points per question

(1) $\begin{cases} 3x - 5y = 8 & \cdots ① \\ -2x + 7y = 2 & \cdots ② \end{cases}$

Check!

① : $3 \times \boxed{} - 5 \times \boxed{} = \boxed{} - \boxed{} =$

The left side of ① matches the right side.

② : $-2 \times \boxed{} + 7 \times \boxed{} = \boxed{} + \boxed{} =$

The left side of ② matches the right side.

Therefore, your answer is correct!

⟨Ans.⟩ _____

(2) $\begin{cases} -8x + 3y = 8 & \cdots ① \\ 9x + 5y = -9 & \cdots ② \end{cases}$

⟨Ans.⟩ _____

(3) $\begin{cases} 5x - 6y = -12 & \cdots ① \\ 4x - 9y = -11 & \cdots ② \end{cases}$

⟨Ans.⟩ _____

Double checking your work takes time, but it's worthwhile!

Simultaneous Linear Equations

Level

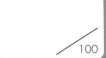

Date   /   /

Name

Score

/100

1 Rewrite each equation into the form $ax + by = c$ **and solve for both variables.**

10 points per question

Don't forget!

In order to solve equations, it may be easier to first rearrange and/or simplify the equations into the form $ax + by = c$.

(1) $\begin{cases} 5y = -2x - 8 & \cdots ① \\ -4y = 3x - 2 & \cdots ② \end{cases}$

① becomes : $\qquad 2x + 5y = -8 \quad \cdots ③$

② becomes : $\boxed{}x - 4y = -2 \quad \cdots ④$

③ × 3 : $\boxed{}x + 15y = -24 \quad \cdots ⑤$

④ × $\boxed{}$: $-6x - 8y = -4 \quad \cdots ⑥$

⑤ + ⑥ : $\boxed{}y = -28$

$y = \boxed{}$

⟨Ans.⟩ _____

(3) $\begin{cases} 7x = 3y - 2 \\ 5x = -26 - 4y \end{cases}$

⟨Ans.⟩ _____

(2) $\begin{cases} -2y = -3x + 9 \\ 7y - 1 = 4x \end{cases}$

⟨Ans.⟩ _____

(4) $\begin{cases} 3y = 6x + 3 \\ -9x - 9 = -5y \end{cases}$

⟨Ans.⟩ _____

2 **Solve each equation.**

15 points per question

(1) $\begin{cases} 4x+2y=x+8y+18 & \cdots ① \\ 5x-13y=6x-9y & \cdots ② \end{cases}$

① becomes： $3x-\boxed{}y=18 \cdots ③$

② becomes： $-x-\boxed{}y=0 \cdots ④$

(3) $\begin{cases} -2(3x+2)=4-y \\ 5(x-3y)=2(6x-5y)-3 \end{cases}$

⟨Ans.⟩ _____

⟨Ans.⟩ _____

(2) $\begin{cases} 3(x+4y)=x+7y+2 & \cdots ① \\ -9x-10y+28=-12x-2y & \cdots ② \end{cases}$

① becomes： $\boxed{}x+5y=2 \cdots ③$

② becomes： $\boxed{}x-\boxed{}y=-28 \cdots ④$

(4) $\begin{cases} -(3-2x)=4(x-y) \\ 2(x-4+y)=-3(y+3-2x) \end{cases}$

⟨Ans.⟩ _____

⟨Ans.⟩ _____

Remove parentheses first.

You are getting good at this!

36 Simultaneous Linear Equations ★★★

Level

Score

/100

Date / /

Name

1 Solve each equation.

15 points per question

(1) $\begin{cases} 4x + \dfrac{2}{3}y = -2 & \cdots\text{①} \\ 5x = 7 - 4y & \cdots\text{②} \end{cases}$

Remove the denominator from ①:

①×3: $12x + \boxed{}\,y = \boxed{}$ ···③

Rearrange the terms in ②:

$5x + \boxed{}\,y = 7$ ···④

③×2:

⟨**Ans.**⟩ _____

(2) $\begin{cases} \dfrac{5}{6}x + \dfrac{1}{4}y = \dfrac{1}{3} & \cdots\text{①} \\ 7x - 4 = -2y - 2 & \cdots\text{②} \end{cases}$

Remove the denominator from ①:

①×12:

⟨**Ans.**⟩ _____

(3) $\begin{cases} \dfrac{2}{5}x - 1 = \dfrac{1}{2}y & \cdots\text{①} \\ \dfrac{1}{4}(x - 2y) = \dfrac{2}{3}(x - y) - \dfrac{7}{4} & \cdots\text{②} \end{cases}$

Remove the denominator from ①:

①×10: $4x - 10 = \boxed{}\,y$

∴ $4x - \boxed{}\,y = 10$ ···③

Remove the denominator from ②:

②×12: $3(x - 2y) = \boxed{}(x - y) - 21$

∴ $-5x + 2y = \boxed{}$ ···④

∴ means "therefore."

⟨**Ans.**⟩ _____

(4) $\begin{cases} -\dfrac{1}{3}(4x - 5y) + 1 = \dfrac{1}{2}(3y - 2x) \\ \dfrac{2}{5}(4 - 2x) + \dfrac{3}{5}y = -\dfrac{1}{2}x + 1 \end{cases}$

⟨**Ans.**⟩ _____

2 **Solve each equation.**

(1) $\begin{cases} \dfrac{x-y}{3} + \dfrac{2x-y}{4} = 2 & \cdots ① \\ \dfrac{x-y}{2} + \dfrac{x+y}{3} = \dfrac{7}{6} & \cdots ② \end{cases}$

Remove the denominator from ①

and ② by :

① × 12 : $4(x-y) + \boxed{}(2x-y) = 24$

② × $\boxed{}$: $3(x-y) + \boxed{}(x+y) = \boxed{}$

$\therefore \begin{cases} \boxed{}x - \boxed{}y = 24 \\ \boxed{}x - y = \boxed{} \end{cases}$

Use ∴ when completely rearranging an equation.

〈Ans.〉 _____

(2) $\begin{cases} \dfrac{2x-3y}{4} - \dfrac{x-y}{2} = \dfrac{x+1}{5} & \cdots ① \\ -\dfrac{x+2y+1}{3} + 1 = \dfrac{-3x-2y-1}{2} & \cdots ② \end{cases}$

Remove the denominator from ①

and ② by :

〈Ans.〉 _____

You went above and beyond!

37 Simultaneous Linear Equations ★★★

Level

Score

/100

Date / /

Name

1 Solve each equation.

14 points per question

Example

$$\begin{cases} y = 2x - 1 & \cdots① \\ 4x - 3y = -7 & \cdots② \end{cases}$$

Step 1: Number each equation.

Substitute ① into ② :

$$4x - 3(2x - 1) = -7$$
$$4x - 6x + 3 = -7$$
$$-2x = -10$$
$$x = 5$$

Step 2: Substitute one equation into the other to remove a variable.

Substitute this into ① :

$$y = 10 - 1$$
$$= 9$$

Step 3: Substitute the value of the variable into either original equation (usually the simpler one) to solve for the other variable.

⟨**Ans.**⟩ $(x, y) = (5, 9)$

Step 4: Write your answer.

The **substitution method** allows you to substitute one equation into the other equation in order to eliminate one variable and solve for the remaining variable.

(1) $$\begin{cases} y = 3x - 2 \\ 2x - y = -1 \end{cases}$$

(2) $$\begin{cases} x = y + 4 \\ 2x + 3y = -2 \end{cases}$$

⟨Ans.⟩

⟨Ans.⟩

2 **Solve each equation by using the substitution method.**

(1) $\begin{cases} 4x+5y=10 & \cdots① \\ x+3y=-1 & \cdots② \end{cases}$

        Rewrite ② : $x = \boxed{}\,y-1$ $\cdots③$

        Substitute ③ into ① :

        $4\left(\boxed{}\,y-1\right)+5y=10$

                      $\boxed{}\,y=14$

                      $y=\boxed{}$

        Substitute this into ③ :

           $x=\boxed{}-1$

           $x=$

      ⟨Ans.⟩ $(x,y)=(\boxed{}, \boxed{})$

(2) $\begin{cases} 3x-5y=-12 \\ x=y \end{cases}$

      ⟨Ans.⟩ _____

(3) $\begin{cases} 5x+3y=8 & \cdots① \\ -4x+7y=3 & \cdots② \end{cases}$

        Rewrite ① : $y = \dfrac{\boxed{}}{3}$ $\cdots③$

        Substitute ③ into ② :

      ⟨Ans.⟩ _____

(4) $\begin{cases} 7x+2y=-6 \\ 8x+5y=4 \end{cases}$

      ⟨Ans.⟩ _____

You can always go back and check your answers by plugging the value of each variable into both equations.

There is no substitute for hard work and practice!

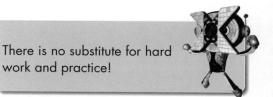

Simultaneous Linear Equations ★★★

Level

Score

/100

Date / /

Name

1 Solve each equation by using the substitution method.

10 points per question

(1) $\begin{cases} 2x = -3y + 7 & \cdots ① \\ 3x - 4y = 2 & \cdots ② \end{cases}$

Rewrite ① : $x = \dfrac{-3y + 7}{\boxed{}}$ $\cdots ③$

Substitute ③ into ② :

$3\left(\dfrac{-3y + 7}{\boxed{}}\right) - 4y = 2$

$3(-3y + 7) - 8y = \boxed{}$

$-17y = \boxed{}$

$y = \boxed{}$

Substitute this into ③ :

$x = \dfrac{\boxed{} + 7}{\boxed{}}$

$x =$

⟨**Ans.**⟩ _____

(2) $\begin{cases} 5x = 7y + 1 \\ 3y = 4x - 6 \end{cases}$

⟨**Ans.**⟩ _____

(3) $\begin{cases} 4x + 5y = -6 & \cdots ① \\ -2x - 3y = 2 & \cdots ② \end{cases}$

Rewrite ② : $x = \dfrac{\boxed{}}{2}$ $\cdots ③$

⟨**Ans.**⟩ _____

(4) $\begin{cases} 4y = 1 - 3x \\ -5x = -3 + 6y \end{cases}$

⟨**Ans.**⟩ _____

© *Kumon Publishing Co., Ltd.*

2 **Use the following methods to solve the equations** $\begin{cases} -3x - 4y = -20 \\ 2x + 5y = 18 \end{cases}$. 15 points per question

（1） Addition/subtraction method

（2） Substitution method

〈Ans.〉 _____

〈Ans.〉 _____

3 **Use the following methods to solve the equations** $\begin{cases} 8x - y = -13 \\ 5x + 3y = 10 \end{cases}$. 15 points per question

（1） Addition/subtraction method

（2） Substitution method

〈Ans.〉 _____

〈Ans.〉 _____

You can solve equations with a few methods now!

77

Simultaneous Linear Equations

Level ★★★

Score
/100

Date   /   /

Name

---

1 Use the following methods to solve the equations $\begin{cases} 5x + 3y = 5 \\ 4x + y = -3 \end{cases}$.

10 points per question

(1)   Addition/subtraction method

(2)   Substitution method

⟨Ans.⟩ _____

⟨Ans.⟩ _____

---

2 Use the following methods to solve the equations $\begin{cases} x - 5y = 8 \\ 7x + 3y = -20 \end{cases}$.

10 points per question

(1)   Addition/subtraction method

(2)   Substitution method

⟨Ans.⟩ _____

⟨Ans.⟩ _____

© Kumon Publishing Co., Ltd.

3 Use the following methods to solve the equations $\begin{cases} 3x - 4y = -2 \\ 2x - 5y = 8 \end{cases}$.

15 points per question

(1)　Addition/subtraction method

(2)　Substitution method

〈Ans.〉

〈Ans.〉

4 Use the following methods to solve the equations $\begin{cases} 3x - 4y = -7 \\ -5x + 6y = 12 \end{cases}$.

15 points per question

(1)　Addition/subtraction method

(2)　Substitution method

〈Ans.〉

〈Ans.〉

When you have completed each exercise, compare the methods and think about which method was more efficient.

You're good at every method!

Word Problems with Simultaneous Linear Equations

Score

/100

Date / /

Name

---

1 Write each word problem as equations, and then solve.   20 points per question

(1) There are two numbers: x and y. The sum of the numbers is 15. The sum of 3 times x and 2 times y is 34. What are the values of x and y?

$$\begin{cases} x+y=15 \\ 3x+\boxed{}y=\boxed{} \end{cases}$$

⟨Ans.⟩ $x=\boxed{}$
$y=\boxed{}$

(2) There are two numbers: a and b. The difference of the numbers is 3. The sum of 4 times a and 5 times b is 39. What are the values of a and b? (Assume that a is larger than b.)

⟨Ans.⟩ $a=\boxed{}$
$b=\boxed{}$

(3) There are two numbers: f and g. 2 times the sum of f and g is 20. $\frac{2}{3}$ of f minus $\frac{1}{4}$ of g is 3. What are the values of f and g?

$$\begin{cases} 2(f+g)=\boxed{} \\ \dfrac{2}{3}f-\boxed{}g=3 \end{cases}$$

⟨Ans.⟩ $f=\boxed{}$
$g=\boxed{}$

2 Write each word problem as equations, and then solve.

20 points per question

(1) If we fill a wheelbarrow with dirt, it weighs 60 pounds. When we only fill the wheelbarrow $\frac{1}{4}$ full, it weighs 33 pounds. How much does the empty wheelbarrow weigh?

Let x be the weight of the empty wheelbarrow

and y be the weight of the dirt in a full wheelbarrow.

$$\begin{cases} x + y = \boxed{} \\ x + \boxed{}\,y = 33 \end{cases}$$

⟨Ans.⟩ _____

(2) If we fill a cart $\frac{1}{2}$ full with books, it weighs 50 pounds. When we fill a cart only $\frac{1}{3}$ full, it weighs 40 pounds. How much does the empty cart weigh?

Let x be the weight of the empty cart

and y be the weight of the books in a full cart.

⟨Ans.⟩ _____

Hooray for you!

Word Problems with Simultaneous Linear Equations

41

Level
★★★

Score

/100

Date / /

Name

1 **Write the word problem as equations, and then solve.**

50 points for completion

Angela had a bracelet that was 3 parts gold and 1 part silver. Christina had another bracelet that was 1 part gold and 2 parts silver. Angela and Christina melted and combined both bracelets into a new necklace. The new necklace contains 6 ounces of gold and 7 ounces of silver. How many ounces did Angela's bracelet weigh, and how many ounces did Christina's bracelet weigh?

If Angela's bracelet had 3 parts gold and 1 part silver, then there were a total of $3 + 1 = \boxed{}$ parts.

If Christina's bracelet had 1 part gold and 2 parts silver, then there were a total of $\boxed{}$ parts.

If x represents the total weight of Angela's bracelet,

then $\dfrac{3}{\boxed{}} x$ represents the weight of the gold in Angela's bracelet

and $\boxed{}$ represents the weight of the silver in Angela's bracelet.

If y represents the total weight of Christina's bracelet,

then $\boxed{}$ represents the weight of the gold in Christina's bracelet

and $\boxed{}$ represents the weight of the silver in Christina's bracelet.

$$\begin{cases} \dfrac{3}{\boxed{}}x + \dfrac{1}{3}y = 6 \\ \boxed{}\,x + \boxed{}\,y = 7 \end{cases}$$

⟨**Ans.**⟩ Angela's bracelet weighed _____ ounces.

Christina's bracelet weighed _____ ounces.

2 **Write the word problem as equations, and then solve.**

50 points for completion

Suzette had a ring that was 1 part steel and 5 parts iron. Chayce had a ring that was 5 parts steel and 3 parts iron. Suzette and Chayce melted and combined both rings into a new ring. The new ring contains 7 grams of steel and 13 grams of iron. How many grams did Suzette's ring weigh, and how many grams did Chayce's ring weigh?

⟨**Ans.**⟩ Suzette's ring weighed   grams.

Chayce's ring weighed   grams.

You thought that through! Good job!

Review

42

Date ___/___/___   Name

Level ★★

Score ___/100

---

1   **Determine the value of each expression when** $a = 3$, $b = -\dfrac{1}{4}$, **and** $c = -2$.

5 points per question

(1) $a(b - c) =$

(2) $(a - b) \div (b + c) =$

(3) $-c^3 - b^2 =$

(4) $\dfrac{a^3}{bc^4} =$

---

2   **Simplify each expression.**

5 points per question

(1) $2x^2 + 3y - 6x^2 =$

(2) $(-2x + y - 3z) + (x - 5y + 2z)$

    $=$

(3) $\left(f - \dfrac{3}{2}g\right) - \left(-2g + \dfrac{1}{4}h\right)$

    $=$

(4) $2(a - b) - 3(-3a - b) =$

(5) $\dfrac{3x - 7}{5} - \dfrac{2x + 9}{5} =$

(6) $\dfrac{2x - 3}{4} - \dfrac{-9x + 7}{6} =$

3 Solve each equation.

(1) $3x + 2 = 14$

(3) $7x = -x - 10$

(2) $2x + 4 = \dfrac{1}{2}$

(4) $\dfrac{1}{2}x = -3 + \dfrac{2}{3}x$

4 Solve each equation, and check your answer.

(1) $2x + 6 = -x - 18$ Check!

(2) $\dfrac{3}{4}x - 5 = \dfrac{3}{2}x - 11$ Check!

5 Answer the word problem.

Tony has 26 marbles, and Lisa has 7 marbles. If Tony gives Lisa x marbles, he will have 3 more marbles than Lisa. How many marbles did Tony give Lisa?

You're almost at the finish line!

⟨Ans.⟩

1   **Solve the equations by using either the addition or subtraction method.**

10 points per question

(1) $\begin{cases} 2x - 3y = -8 \\ 6x + 3y = 0 \end{cases}$

(2) $\begin{cases} 2x - 5y = -2 \\ 2x + 3y = 14 \end{cases}$

⟨Ans.⟩ _____

⟨Ans.⟩ _____

2   **Solve the equations by using the substitution method.**

10 points per question

(1) $\begin{cases} 3x + y = 11 \\ y = -2x + 8 \end{cases}$

(2) $\begin{cases} x = y - 2 \\ 2x - 3y = -9 \end{cases}$

⟨Ans.⟩ _____

⟨Ans.⟩ _____

3 Solve each equation.

10 points per question

(1) $\begin{cases} 2x - y = -5 \\ 3x + 2y = -4 \end{cases}$

(3) $\begin{cases} 2x + 5y = -4 \\ 3x - 4y = 17 \end{cases}$

⟨Ans.⟩ _____

⟨Ans.⟩ _____

(2) $\begin{cases} -3x + 7y = -8 \\ 6x + 5y = -3 \end{cases}$

(4) $\begin{cases} 9x + 2y = -12 \\ -6x - 5y = 19 \end{cases}$

⟨Ans.⟩ _____

⟨Ans.⟩ _____

4 Answer the word problem.

20 points for completion

Paul fills a jar with sand. When the jar is completely filled, it weighs 100 grams. When the jar is $\frac{3}{5}$ full of sand, it weighs only 76 grams. How much does the empty jar weigh?

⟨Ans.⟩ _____

Congratulations on completing
Algebra Workbook I!

87

① Fractions Review   pp 2, 3

①
(1) $\frac{11}{18}$ (6) $\frac{3}{40}$

(2) $\frac{7}{20}$ (7) 1

(3) $1\frac{3}{20}$ (8) $3\frac{5}{8}$

(4) $1\frac{7}{12}$ (9) $8\frac{7}{24}$

(5) $2\frac{17}{40}$ (10) $9\frac{1}{8}$

②
(1) $\frac{1}{32}$ (6) $1\frac{1}{2}$

(2) $1\frac{3}{8}$ (7) $\frac{18}{25}$

(3) $\frac{2}{5}$ (8) $\frac{2}{7}$

(4) $1\frac{3}{4}$ (9) $\frac{1}{5}$

(5) $2\frac{1}{4}$ (10) 6

② Exponents Review   pp 4, 5

①
(1) 8 (6) $2^{\boxed{5}} = 32$

(2) $1\frac{11}{25}$ (7) $\frac{16}{81}$

(3) $\frac{1}{36}$ (8) $\frac{9}{64}$

(4) $\frac{24}{25}$ (9) $\frac{64}{81}$

(5) $60\frac{3}{4}$ (10) $3\frac{3}{5}$

②
(1) $2^{\boxed{4}} = 16$ (6) $\frac{1}{36}$

(2) 243 (7) $\frac{27}{125}$

(3) $\frac{16}{81}$ (8) $\frac{1}{16}$

(4) 125 (9) 64

(5) $\frac{27}{64}$ (10) $2\frac{1}{4}$

③ Order of Operations Review   pp 6, 7

①
(1) $6\frac{1}{2}$ (5) $7\frac{3}{4}$

(2) $1\frac{1}{4}$ (6) $13\frac{1}{2}$

(3) $\frac{3}{8}$ (7) $8\frac{2}{3}$

(4) 10 (8) $5\frac{1}{6}$

②
(1) $1\frac{7}{12}$ (4) $2\frac{5}{6}$

(2) $1\frac{24}{25}$ (5) $9\frac{71}{72}$

(3) $\frac{24}{29}$ (6) $\frac{19}{72}$

④ Negative Numbers Review   pp 8, 9

①
(1) -4 (8) $-2\frac{3}{20}$

(2) 6 (9) $-1\frac{1}{2}$

(3) 13 (10) $-3\frac{1}{2}$

(4) $-\frac{3}{4}$ (11) $1\frac{1}{2}$

(5) $6\frac{3}{5}$ (12) $4\frac{5}{6}$

(6) $-\frac{2}{3}$ (13) $-9\frac{5}{6}$

(7) $-2\frac{11}{12}$ (14) $-4\frac{5}{6}$

②
(1) -12 (7) $-\frac{1}{40}$

(2) 32 (8) $-1\frac{1}{3}$

(3) $-\frac{3}{10}$ (9) $4\frac{1}{2}$

(4) -6 (10) $-1\frac{1}{2}$

(5) -6 (11) $\frac{3}{4}$

(6) 2 (12) $-2\frac{7}{10}$

⑤ Negative Numbers Review   pp 10, 11

①
(1) -32 (6) $-\frac{1}{4}$

(2) -108 (7) $\frac{1}{4}$

(3) 36 (8) 108

(4) $\frac{1}{8}$ (9) $-\frac{40}{81}$

(5) -4 (10) $-8\frac{1}{3}$

②
(1) $-\frac{2}{3}$ (6) $-1\frac{1}{8}$

(2) $-\frac{11}{14}$ (7) $-\frac{9}{32}$

(3) -48 (8) $30\frac{1}{9}$

(4) $-\frac{3}{40}$ (9) $-\frac{2}{5}$

(5) $-\frac{7}{15}$ (10) $\frac{16}{25}$

⑥ Values of Algebraic Expressions Review   pp 12, 13

①
(1) -12 (6) -3

(2) $2\frac{1}{2}$ (7) 24

(3) -6 (8) $5\frac{1}{2}$

(4) 10 (9) $23\frac{1}{6}$

(5) -21 (10) $-3\frac{3}{4}$

②
(1) $3\frac{1}{3}$ (6) 2

(2) $1\frac{1}{12}$ (7) $\frac{5}{6}$

(3) $5\frac{5}{6}$ (8) $-4\frac{1}{6}$

(4) $\frac{1}{6}$ (9) $-\frac{7}{20}$

(5) -9 (10) $3\frac{1}{2}$

7 Values of Algebraic Expressions Review pp 14, 15

1 (1) -32 (6) -5
(2) 8 (7) -5
(3) -1 (8) 24
(4) $\dfrac{3}{4}$ (9) -36
(5) $-1\dfrac{2}{5}$ (10) -36

2 (1) $2\dfrac{19}{20}$ (6) $\dfrac{15}{64}$
(2) -7 (7) 40
(3) $-\dfrac{1}{24}$ (8) -4
(4) $13\dfrac{1}{3}$ (9) $\dfrac{7}{8}$
(5) $-\dfrac{3}{16}$ (10) $\dfrac{7}{8}$

8 Values of Algebraic Expressions Review pp 16, 17

1 (1) -4 (5) $5\dfrac{1}{4}$
(2) $-2\dfrac{1}{2}$ (6) -7
(3) $-1\dfrac{5}{12}$ (7) $-1\dfrac{2}{3}$
(4) $-\dfrac{1}{2}$ (8) -7

2 (1) -4 (4) $2\dfrac{7}{8}$
(2) -3 (5) 3
(3) -3 (6) 3

9 Values of Algebraic Expressions Review pp 18, 19

1 (1) 9 (5) $31\dfrac{1}{4}$
(2) $\dfrac{2}{5}$ (6) $24\dfrac{1}{2}$
(3) $-\dfrac{5}{6}$ (7) 1
(4) $-\dfrac{5}{6}$ (8) $-1\dfrac{11}{16}$

2 (1) $-\dfrac{2}{3}$ (4) $\dfrac{5}{6}$
(2) $2\dfrac{1}{2}$ (5) $-9\dfrac{2}{27}$
(3) $\dfrac{1}{6}$ (6) $-2\dfrac{4}{9}$

10 Simplifying Algebraic Expressions pp 20, 21

1 (1) $5\boxed{x}$ (6) $5b$
(2) $15y$ (7) $3h$
(3) $8d$ (8) x
(4) $4x$ (9) $3y$
(5) $8z$ (10) $22w$

2 (1) $14x$ (9) $\dfrac{5}{9}z$
(2) $3y$ (10) $-\dfrac{1}{2}f$
(3) $5a+\boxed{4a}=9a$ (11) $\dfrac{\boxed{6}}{3}x+\dfrac{1}{3}x=\dfrac{\boxed{7}}{3}x$
(4) $-13h$ (12) $\dfrac{11}{4}y$
(5) $7a$ (13) $\dfrac{7}{3}b$
(6) $5j$ (14) m
(7) $-5j$ (15) $\boxed{3}xy$
(8) $\dfrac{2}{7}w$

11 Simplifying Algebraic Expressions pp 22, 23

1 (1) $\boxed{6}x+5$ (8) $\boxed{-4}a+\boxed{7}b$
(2) $14y-3$ (9) $a-\dfrac{4}{3}b$
(3) $2h-6$ (10) $5w+y$
(4) $\dfrac{23}{12}bc+\dfrac{2}{7}$ (11) $-\dfrac{13}{6}a-x$
(5) $-\dfrac{5}{4}ab-\dfrac{5}{6}$ (12) $\dfrac{7}{3}fg+2gh$
(6) $-7x+5$ (13) $\dfrac{19}{8}c-\dfrac{3}{10}h$
(7) $3x-11$

2 (1) $\boxed{7}x^2-8x$ (7) $\boxed{8}a+\boxed{10}b+\boxed{13}c$
(2) $3x^2-7x$ (8) $-2x-\dfrac{17}{4}y+\dfrac{3}{2}$
(3) $\dfrac{9}{4}y^2-\dfrac{3}{2}$ (9) $10b-6c^2+\dfrac{5}{6}$
(4) $-\dfrac{7}{3}a^2+\dfrac{1}{2}bc$ (10) $\boxed{9}x^2+\boxed{4}$
(5) $-s+6v^3$ (11) $3y^2+12$
(6) $5n^2-\dfrac{10}{3}w$ (12) $-\dfrac{11}{4}abc+\dfrac{11}{2}be$

(12) Simplifying Algebraic Expressions pp 24, 25

1 (1) $4x - 2y$

(2) $-2a - 8b$

(3) $g + 3h$

(4) $-2m + 6n$

(5) $-\dfrac{3}{2}a + \dfrac{7}{2}b$

(6) $-\dfrac{7}{6}k + \dfrac{3}{4}l$

(7) $-x + \dfrac{5}{2}$

(8) $-\dfrac{5}{6}s - \dfrac{17}{20}t$

2 (1) $2a - 5b + 10c$

(2) $3x - 5y + z$

(3) $2a + \dfrac{5}{4}c$

(4) $-\dfrac{11}{4}b + \dfrac{9}{4}c^2 - \dfrac{11}{12}de$

(5) $5a + 5b - 3c + 2$

(6) $\dfrac{1}{4}e - \dfrac{3}{2}f + \dfrac{3}{2}g - \dfrac{1}{6}h$

---

(13) Simplifying Algebraic Expressions pp 26, 27

1 (1) $x \boxed{-} y \boxed{+} z$

(2) $a \boxed{-} b \boxed{+} c$

(3) $-g \boxed{+} h \boxed{+} i$

(4) $-l \boxed{+} m \boxed{-} n$

(5) $x \boxed{-} y \boxed{-} z$

(6) $a \boxed{+} b \boxed{+} c$

(7) $g \boxed{-} h \boxed{-} 2i$

(8) $l \boxed{-} 3m \boxed{-} n$

2 (1) $5x + 3$

(2) $13x + 3$

(3) $13x - 3$

(4) $-8x + 7$

(5) $-2x - 7$

(6) $-8x - 7$

(7) $\dfrac{7}{2}a + \dfrac{5}{2}$

(8) $-\dfrac{7}{2}a - \dfrac{5}{2}$

(9) $a - \dfrac{5}{2}b$

(10) $\dfrac{3}{4}x - \dfrac{4}{3}y$

---

(14) Simplifying Algebraic Expressions pp 28, 29

1 (1) $2x + 2y$

(2) $-2x - 2y$

(3) $8x - 2y$

(4) $-2x + 2y$

(5) $-8x - 10y$

(6) $-2a - b$

(7) $-2a + b$

(8) $-4a + 3b$

(9) $2a - 3b$

(10) $-4a + b$

2 (1) $6a - 12b$

(2) $-2a + 6b$

(3) $2a + 12b$

(4) $-6a + 6b$

(5) $-x + 4y$

(6) $-\dfrac{3}{4}f + \dfrac{8}{3}g$

(7) $-\dfrac{14}{15}y - \dfrac{3}{20}z$

(8) $\dfrac{13}{12}m - \dfrac{19}{12}n$

(9) $-\dfrac{15}{4}a + \dfrac{29}{10}b$

(10) $\dfrac{22}{15}x + \dfrac{17}{8}y$

---

(15) Simplifying Algebraic Expressions pp 30, 31

1 (1) $4x + 6y$

(2) $6x - 4y$

(3) $6x - 5y$

(4) $-\dfrac{4}{7}x - 3y$

(5) $2b + 13c$

(6) $6x - 15y$

(7) $\boxed{5}t$

(8) $\boxed{2}a$

2 (1) $10a + 8b + 5c$

(2) $11a + 4b - 5c$

(3) $-7x - 2y + z$

(4) $-\dfrac{3}{2}x^2 - \dfrac{2}{15}x - \dfrac{11}{4}$

(5) $\boxed{0}$

(6) $13x + 3y + \boxed{8}z$

(7) $2x + 4y - 9z$

(8) $a - 3b + 8c$

(9) $\dfrac{1}{6}x^2 - \dfrac{1}{3}xy + \dfrac{9}{20}y^2$

(10) $\dfrac{7}{12}a^2 - \dfrac{4}{3}a + 5$

---

(16) Simplifying Algebraic Expressions pp 32, 33

1 (1) $cd + \boxed{ce}$

(2) $\boxed{xy} + \boxed{xz}$

(3) $\boxed{ab} - \boxed{ac}$

(4) $2x - 2y$

(5) $3s - 3t$

(6) $3x + \boxed{2}x + 2y = 5x + 2y$

(7) $x + 3y$

(8) $-6x + 6y$

(9) $-\dfrac{5}{2}a + \dfrac{2}{3}b$

(10) $\dfrac{5}{4}m + \dfrac{1}{3}n$

② (1) $6x - 12y + \boxed{6}x - \boxed{10}y = 12x - 22y$

(2) $4x + 2y$

(3) $8x - 11y$

(4) $\dfrac{7}{3}a - b$

(5) $-\dfrac{19}{6}a^2 - a$

(6) $10x^2 + 25$

(7) $14x + 8y$

(8) $-x - 13y$

(9) $-7g + 13h$

(10) $-\dfrac{15}{2}a + \dfrac{15}{2}b$

(11) $-\dfrac{23}{10}c + \dfrac{1}{12}d$

(12) $\dfrac{9}{2}x + \dfrac{13}{2}y - \dfrac{13}{6}z$

17 Simplifying Algebraic Expressions pp 34, 35

① (1) $\dfrac{(\boxed{6x+9}) + (\boxed{x-6})}{4} = \dfrac{7x+3}{4}$

(2) $\dfrac{-3x+7}{5}$

(3) $\dfrac{(\boxed{5x+11}) - (\boxed{3x+4})}{3} = \dfrac{2x+7}{3}$

(4) $\dfrac{x+14}{6}$

(5) $\dfrac{-(\boxed{4x+7}) - (\boxed{3x-2})}{2} = \dfrac{-7x-5}{2}$

(6) $\dfrac{-14x+12}{7}$

(7) $\dfrac{-6x+5}{4}$

(8) $\dfrac{x-8}{8}$

② (1) $\boxed{4}$, $\boxed{3}$, $\dfrac{20x+27}{12}$

(2) $\dfrac{18x-11}{6}$

(3) $\dfrac{8x-11}{10}$

(4) $\boxed{3}$, $\boxed{2}$, $\dfrac{7x+19}{12}$

(5) $\dfrac{12x-7}{18}$

(6) $\dfrac{7x+1}{4}$

(7) $\dfrac{-10x+1}{6}$

(8) $\dfrac{x-5}{9}$

(9) $\dfrac{5x+2}{3}$

(10) $\dfrac{-x+2}{5}$

18 Solving Equations pp 36, 37

① (1) $x = 7 + \boxed{2}$

$x = 9$

(2) $x = -8 + \boxed{5}$

$x = \boxed{-3}$

(3) $x = -3$

(4) $x = 8$

(5) $x = -7$

(6) $x = -6$

(7) $x = 15$

(8) $x = 3$

(9) $x = 6$

(10) $x = 0$

② (1) $x = 4$

(2) $x = -4$

(3) $x = -5$

(4) $x = 0$

(5) $x = 2.6$

(6) $x = 1.5$

(7) $x = 4.1$

(8) $x = -3.5$

(9) $x = -\dfrac{4}{3}$

(10) $x = \dfrac{9}{4}$

(11) $x = -\dfrac{13}{2}$

(12) $x = \dfrac{5}{6}$

(13) $x = -\dfrac{1}{20}$

(14) $x = \dfrac{13}{4}$

(15) $x = \dfrac{3}{4}$

(16) $x = \dfrac{1}{2}$

19 Solving Equations pp 38, 39

① (1) $x = 30 \times \dfrac{1}{\boxed{5}}$

$x = 6$

(2) $x = 7$

(3) $x = 6 \times (-\boxed{1})$

$x = -6$

(4) $x = 18 \times \left(-\dfrac{1}{\boxed{3}}\right)$

$x = -6$

(5) $x = -\dfrac{7}{2}$

(6) $x = 12$

(7) $x = -10$

(8) $x = \dfrac{3}{4}$

(9) $x = -18$

(10) $x = \dfrac{9}{2}$

② (1) $x = -8$

(2) $x = 15$

(3) $x = 12$

(4) $x = \dfrac{15}{2}$

(5) $x = -\dfrac{28}{3}$

(6) $x = -\dfrac{3}{16}$

(7) $x = -\dfrac{3}{8}$

(8) $x = \dfrac{6}{5}$

(9) $x = \dfrac{33}{28}$

(10) $x = -\dfrac{5}{14}$

20 Solving Equations

1 (1) $3x = 10 + \boxed{2}$

$\qquad 3x = \boxed{12}$

$\qquad x = \boxed{12} \times \dfrac{1}{\boxed{3}}$

$\qquad x = 4$

(2) $x = -3$

(3) $x = \dfrac{3}{4}$

(4) $x = \dfrac{7}{3}$

(5) $x = -1$

(6) $\dfrac{3}{5}x = 2 + \boxed{4}$

$\qquad \dfrac{3}{5}x = \boxed{6}$

$\qquad x = \boxed{6} \times \dfrac{5}{\boxed{3}}$

$\qquad x = 10$

2 (1) $x = 2$

(2) $x = -\dfrac{1}{3}$

(3) $x = -\dfrac{5}{6}$

(4) $x = \dfrac{4}{5}$

(5) $x = -2$

(6) $x = -\dfrac{1}{6}$

(7) $x = -24$

(8) $x = \dfrac{12}{5}$

21 Solving Equations

1 (1) $2x + 6x - \boxed{7} = 9$

$\qquad \boxed{8}x - \boxed{7} = 9$

$\qquad\qquad x = 2$

(2) $x = -4$

(3) $x = -6$

(4) $x = -3$

2 (1) $x = -\dfrac{1}{5}$

(2) $x = -\dfrac{2}{13}$

(3) $x = \dfrac{7}{3}$

(4) $x = -18$

(5) $x = -\dfrac{5}{6}$

(6) $x = \dfrac{8}{7}$

22 Solving Equations

1 (1) $\boxed{-}$, $\boxed{+}$, $x = 4$

(2) $\boxed{-}$, $\boxed{-}$, $x = -3$

(3) $x = -1$

(4) $x = 2$

(5) $x = -1$

(6) $x = -2$

2 (1) $x = 3$

(2) $x = -4$

(3) $x = -\dfrac{9}{2}$

(4) $x = 2$

(5) $x = \dfrac{1}{2}$

(6) $x = \dfrac{2}{3}$

(7) $x = \dfrac{16}{3}$

(8) $x = -\dfrac{5}{8}$

23 Solving Equations

1 (1) $2x + \boxed{3x} = 4 + \boxed{6}$

$\qquad x = 2$

Left side $= 2 \times \boxed{2} - 6 = -2$

Right side $= -3 \times \boxed{2} + 4 = -6 + 4 = -2$

(2) $x = -2$

Left side $= 4 \times (-2) + 7 = -1$

Right side $= 9 + 5 \times (-2) = -1$

2 (1) $x = -2$

(2) $x = 6$

(3) $x = \dfrac{4}{3}$

(4) $x = \dfrac{3}{25}$

24 Solving Equations

1 (1) $2x - \boxed{10} = -4$

$\qquad x = 3$

(2) $x = -8$

(3) $x = 1$

(4) $x = 2$

(5) $x = 8$

(6) $x = \dfrac{9}{2}$

2 (1) $4 + \boxed{2}x + \boxed{8} = -x$

$\qquad x = -4$

(2) $x = -1$

(3) $x = -3$

(4) $x = -\dfrac{4}{3}$

(5) $x = \dfrac{8}{5}$

(6) $x = \dfrac{8}{3}$

(7) $x = -\dfrac{8}{15}$

(8) $x = -18$

25 Solving Equations

1 (1) $\boxed{12}$, $\boxed{12}$, $\boxed{12}$, $x = -2$

(2) $x = 1$

(3) $x = -24$

(4) $x = -\dfrac{60}{11}$

2 (1) $\boxed{20}$, $\boxed{20}$, $x = 4$

(2) $x = 8$

(3) $x = -\dfrac{3}{4}$

(4) $x = -\dfrac{3}{11}$

(5) $x = -8$

(6) $x = -1$

26 Word Problems with Equations

1 (1) $3 + x = \boxed{7}$, $x = 4$

(2) $\frac{1}{2} + x = -3$,

$x = -\frac{7}{2}$ $\left(\text{or} -3\frac{1}{2}\right)$

(3) $\boxed{x} - 2 = 10$, $x = 12$

(4) $x - \frac{1}{2} = \frac{1}{3}$, $x = \frac{5}{6}$

(5) $5 - x = \boxed{4}$, $x = 1$

(6) $x = 6 \times (\boxed{-2})$, $x = -12$

(7) $\frac{3}{4} \times x = 6$, $x = 8$

(8) $\frac{x}{5} = 4$, $x = 20$

(9) $\frac{x}{3} = \frac{2}{5}$, $x = \frac{6}{5}$

(10) $\frac{x}{12} = \frac{2}{3}$, $x = 8$

2 (1) $4 \times x = \boxed{76}$, $x = 19$ **Ans.** 19 yo-yos

(2) $24 \times x = 192$, $x = 8$ **Ans.** 8 ounces

(3) $5 \times 3 + \boxed{4} \times x = 39$, $x = 6$ **Ans.** 6 oranges

(4) $25 \times 3 + 40 \times x = 315$, $x = 6$ **Ans.** 6 grams

27 Word Problems with Equations
pp 54, 55

1 (1) **Ans.** $(\boxed{12} - x)$ pens

(2) **Ans.** $(30 - x)$ crayons

(3) **Ans.** $(x - 2)$ cats

(4) **Ans.** $(25 - x)$ girls

(5) **Ans.** $(5 + x)$ girls

(6) **Ans.** $(x - 5)$ boys

2 Method #1: $\boxed{6}$, $\boxed{6}$, $x = 9$, $24 - 9 = 15$ **Ans.** 9 boys   15 girls

Method #2: $\boxed{6}$, $\boxed{x-6}$, $x = 15$, $24 - 15 = 9$ **Ans.** 9 boys   15 girls

3 Method #1:

If x is the number of bananas,

then $(x + 9)$ is the number of apples.

$x + (x + 9) = 31$

$x = 11$

$31 - 11 = 20$ **Ans.** 20 apples   11 bananas

Method #2:

If x is the number of apples,

then $(x - 9)$ is the number of bananas.

$x + (x - 9) = 31$

$x = 20$

$31 - 20 = 11$ **Ans.** 20 apples   11 bananas

28 Word Problems with Equations
pp 56, 57

1 (1) $\boxed{x}$, $\boxed{8}$, $22 - \boxed{x} = \boxed{8} + x$, $x = 7$ **Ans.** 7 markers

(2) $\boxed{x}$, $20 - \boxed{x} = (6 + x) + \boxed{4}$, $x = 5$ **Ans.** 5 coins

2 (1) $\boxed{x}$, $(12 + \boxed{x}) = \boxed{4} \times 12$, $x = 36$ **Ans.** 36 years

(2) $x + 10 = 3 \times x$, $x = 5$ **Ans.** 5 years old

(2) $\boxed{x}$, $\boxed{x}$, $\boxed{2} \times (10 + \boxed{x}) = 26 + \boxed{x}$,

$x = 6$ **Ans.** 6 years

29 Solving Equations
pp 58, 59

1 (1) $x = b - \boxed{6}$

(2) $x = c - 1$

(3) $x = a + \boxed{8}$

(4) $x = d + 4$

(5) $x = -g + 11$

(6) $x = a - b$

(7) $x = -c - d$

(8) $x = b + \boxed{1}$

(9) $x = -y - 2$

(10) $x = a - 3$

2 (1) $x = \frac{c}{a}$

(2) $x = \frac{b}{-3}$

$x = -\frac{b}{3}$

(3) $x = \frac{g}{f}$

(4) $4x = \boxed{s}$

$x = \frac{s}{4}$

(5) $x = \frac{7}{a}$

(6) $x = \frac{-c + d}{b}$

(7) $cx = d - \boxed{3}$

$x = \frac{d - 3}{c}$

(8) $x = \frac{b + 5}{a}$

(9) $rx = \boxed{s - 2t}$

$x = \frac{s - 2t}{r}$

(10) $x = \frac{-y + 2z}{w}$

30 Solving Equations
pp 60, 61

1 (1) $\boxed{ad}$, $x = \frac{abd - ac}{d}$

(2) $x = -\frac{24}{11}$

(3) $x = -\frac{10}{13}$

(4) $\boxed{bc}$, $x = \frac{cd}{b}$

(5) $x = \frac{bc}{ad}$

(6) $x = -\frac{15fh}{4g}$

2 (1) $x = a + b$

(2) $\boxed{2}$, $x = a + \frac{b}{2}$

(3) $x = \frac{a}{y} - 4$

(4) $x = \frac{c + d}{a} - b$

(5) $x = a + 3b$

(6) $x = -5d - 3$

(7) $x = -\frac{5a - 3}{4}$

(8) $x = -\frac{a(b - d)}{M}$

31 **Simultaneous Linear Equations** pp 62,63

1 (1) $\boxed{6}$, $\boxed{-1}$, $\boxed{-1}$, $\boxed{-12}$, $\boxed{-4}$

Ans. $(x,\ y)=(\boxed{-1},\ \boxed{-4})$

(2) $\boxed{2}$ Ans. $(x,\ y)=(4,\ 5)$

2 (1) Ans. $(x,\ y)=(-1,\ 2)$ (4) Ans. $(x,\ y)=(4,\ 5)$

(2) Ans. $(x,\ y)=(-6,\ 3)$ (5) Ans. $(x,\ y)=(-5,\ 2)$

(3) Ans. $(x,\ y)=(2,\ 1)$

32 **Simultaneous Linear Equations** pp 64,65

1 (1) Ans. $(x,\ y)=(2,\ -1)$ (2) Ans. $(x,\ y)=(-5,\ -3)$

2 (1) $\boxed{+}$ Ans. $(x,\ y)=(-4,\ -1)$

(2)   Ans. $(x,\ y)=\left(\dfrac{1}{2},\ -4\right)$

(3)   Ans. $(x,\ y)=(6,\ 3)$

(4)   Ans. $(x,\ y)=\left(\dfrac{7}{10},\ \dfrac{6}{5}\right)$

33 **Simultaneous Linear Equations** pp 66,67

1 (1) $\boxed{2}$, $\boxed{22}$, $\boxed{27}$, $\boxed{3}$, $\boxed{3}$, $\boxed{5}$ Ans. $(x,\ y)=(\boxed{3},\ \boxed{5})$

(2)   Ans. $(x,\ y)=(2,\ -1)$

2 (1)   Ans. $(x,\ y)=(3,\ 2)$

①: $2\times\boxed{3}+3\times\boxed{2}=\boxed{6}+\boxed{6}=\boxed{12}$

②: $5\times\boxed{3}-\boxed{2}=\boxed{15}-\boxed{2}=\boxed{13}$

(2)   Ans. $(x,\ y)=(1,\ -3)$

34 **Simultaneous Linear Equations** pp 68,69

1 (1) ①×4 : $\begin{cases} 20x+12y=\boxed{-28} & \cdots③ \\ 18x+12y=\boxed{-24} & \cdots④ \end{cases}$

②×$\boxed{3}$:

③−④ : $2x=\boxed{-4}$

$x=\boxed{-2}$

$5\times(\boxed{-2})+3y=-7$

$3y=\boxed{3}$

$y=\boxed{1}$

Ans. $(x,\ y)=(\boxed{-2},\ \boxed{1})$

(2) $\boxed{2}$ Ans. $(x,\ y)=(4,\ 2)$

2 (1)   Ans. $(x,\ y)=(6,\ 2)$

①: $3\times\boxed{6}-5\times\boxed{2}=\boxed{18}-\boxed{10}=8$

②: $-2\times\boxed{6}+7\times\boxed{2}=\boxed{-12}+\boxed{14}=2$

(2)   Ans. $(x,\ y)=(-1,\ 0)$

(3)   Ans. $(x,\ y)=\left(-2,\ \dfrac{1}{3}\right)$

35 **Simultaneous Linear Equations** pp 70,71

1 (1) $\boxed{-3}$, $\boxed{6}$, $\boxed{2}$, $\boxed{7}$, $\boxed{-4}$ Ans. $(x,\ y)=(6,\ -4)$

(2)   Ans. $(x,\ y)=(5,\ 3)$

(3)   Ans. $(x,\ y)=(-2,\ -4)$

(4)   Ans. $(x,\ y)=(4,\ 9)$

2 (1) $\boxed{6}$, $\boxed{4}$ Ans. $(x,\ y)=(4,\ -1)$

(2) $\boxed{2}$, $\boxed{3}$, $\boxed{8}$ Ans. $(x,\ y)=(-4,\ 2)$

(3)   Ans. $(x,\ y)=(-1,\ 2)$

(4)   Ans. $(x,\ y)=\left(\dfrac{19}{6},\ \dfrac{7}{3}\right)$

36 **Simultaneous Linear Equations** pp 72,73

1 (1) $\boxed{2}$, $\boxed{-6}$, $\boxed{4}$ Ans. $(x,\ y)=(-1,\ 3)$

(2)   Ans. $(x,\ y)=(-2,\ 8)$

(3) $\boxed{5}$, $\boxed{5}$, $\boxed{8}$, $\therefore -5x+2y=\boxed{-21}$

Ans. $(x,\ y)=(5,\ 2)$

(4)   Ans. $(x,\ y)=\left(\dfrac{10}{3},\ \dfrac{2}{3}\right)$

2 (1) ①×12 : $4(x-y)+\boxed{3}(2x-y)=24$

②×$\boxed{6}$: $3(x-y)+\boxed{2}(x+y)=\boxed{7}$

$\therefore \begin{cases} \boxed{10}x-\boxed{7}y=24 \\ \boxed{5}x-y=\boxed{7} \end{cases}$

Ans. $(x,\ y)=(1,\ -2)$

(2)   Ans. $(x,\ y)=(-1,\ 0)$

94   © Kumon Publishing Co., Ltd.

37 Simultaneous Linear Equations · pp74,75

1 (1) **Ans.** $(x, y)=(3, 7)$ (2) **Ans.** $(x, y)=(2, -2)$

2 (1) Rewrite ② : $x = \boxed{-3}y-1 \cdots ③$

$$4(\boxed{-3}y-1)+5y=10$$
$$\boxed{-7}y=14$$
$$y=\boxed{-2}$$
$$x=\boxed{6}-1$$

Ans. $(x, y)=(\boxed{5}, \boxed{-2})$

(2) **Ans.** $(x, y)=(6, 6)$

(3) $\boxed{8-5x}$ **Ans.** $(x, y)=(1, 1)$

(4) **Ans.** $(x, y)=(-2, 4)$

38 Simultaneous Linear Equations · pp76,77

1 (1) Rewrite ① : $x = \dfrac{-3y+7}{\boxed{2}} \cdots ③$

$$3\left(\dfrac{-3y+7}{\boxed{2}}\right)-4y=2$$
$$3(-3y+7)-8y=\boxed{4}$$
$$-17y=\boxed{-17}$$
$$y=\boxed{1}$$
$$x=\dfrac{\boxed{-3}+7}{\boxed{2}}$$

Ans. $(x, y)=(2, 1)$

(2) **Ans.** $(x, y)=(3, 2)$

(3) $\boxed{-3y-2}$ **Ans.** $(x, y)=(-4, 2)$

(4) **Ans.** $(x, y)=(3, -2)$

2 (1) **Ans.** $(x, y)=(4, 2)$ (2) **Ans.** $(x, y)=(4, 2)$

3 (1) **Ans.** $(x, y)=(-1, 5)$ (2) **Ans.** $(x, y)=(-1, 5)$

39 Simultaneous Linear Equations · pp78,79

1 (1) **Ans.** $(x, y)=(-2, 5)$ (2) **Ans.** $(x, y)=(-2, 5)$

2 (1) **Ans.** $(x, y)=(-2, -2)$ (2) **Ans.** $(x, y)=(-2, -2)$

3 (1) **Ans.** $(x, y)=(-6, -4)$ (2) **Ans.** $(x, y)=(-6, -4)$

4 (1) **Ans.** $(x, y)=\left(-3, -\dfrac{1}{2}\right)$ (2) **Ans.** $(x, y)=\left(-3, -\dfrac{1}{2}\right)$

40 Word Problems with Simultaneous Linear Equations · pp80,81

1 (1) $\begin{cases} x+y=15 \\ 3x+\boxed{2}y=\boxed{34} \end{cases}$

Ans. $\begin{aligned} x &= \boxed{4} \\ y &= \boxed{11} \end{aligned}$

(2) $\begin{cases} a-b=3 \\ 4a+5b=39 \end{cases}$

Ans. $\begin{aligned} a &= \boxed{6} \\ b &= \boxed{3} \end{aligned}$

(3) $\begin{cases} 2(f+g)=\boxed{20} \\ \dfrac{2}{3}f-\dfrac{1}{\boxed{4}}g=3 \end{cases}$

Ans. $\begin{aligned} f &= \boxed{6} \\ g &= \boxed{4} \end{aligned}$

2 (1) $\begin{cases} x+y=\boxed{60} \\ x+\dfrac{1}{\boxed{4}}y=33 \end{cases}$

Ans. 24 pounds

(2) $\begin{cases} x+\dfrac{1}{2}y=50 \\ x+\dfrac{1}{3}y=40 \end{cases}$

Ans. 20 pounds

41 Word Problems with Simultaneous Linear Equations · pp82,83

1 $\boxed{4}, \boxed{3}, \dfrac{3}{\boxed{4}}, \dfrac{1}{\boxed{4}}x, \dfrac{1}{\boxed{3}}y, \dfrac{2}{\boxed{3}}y$

$\begin{cases} \dfrac{3}{\boxed{4}}x+\dfrac{1}{3}y=6 \\ \dfrac{1}{\boxed{4}}x+\dfrac{2}{\boxed{3}}y=7 \end{cases}$

Ans. Angela's bracelet weighed 4 ounces.
Christina's bracelet weighed 9 ounces.

2 If x represents the total weight of Suzette's ring, and y represents the total weight of Chayce's ring, the equation is the following.

$\begin{cases} \dfrac{1}{6}x+\dfrac{5}{8}y=7 \\ \dfrac{5}{6}x+\dfrac{3}{8}y=13 \end{cases}$

Ans. Suzette's ring weighed 12 grams.
Chayce's ring weighed 8 grams.

42 Review pp 84, 85

1 (1) $5\frac{1}{4}$ (3) $7\frac{15}{16}$

(2) $-1\frac{4}{9}$ (4) $-6\frac{3}{4}$

2 (1) $-4x^2+3y$ (4) $11a+b$

(2) $-x-4y-z$ (5) $\dfrac{x-16}{5}$

(3) $f+\dfrac{1}{2}g-\dfrac{1}{4}h$ (6) $\dfrac{24x-23}{12}$

3 (1) $x=4$ (3) $x=-\dfrac{5}{4}$

(2) $x=-\dfrac{7}{4}$ (4) $x=18$

4 (1) $x=-8$

(2) $x=8$

5 $(26-x)=(7+x)+3, \quad x=8$ **Ans.** 8 marbles

---

43 Review pp 86, 87

1 (1) **Ans.** $(x,\ y)=(-1,\ 2)$ (2) **Ans.** $(x,\ y)=(4,\ 2)$

2 (1) **Ans.** $(x,\ y)=(3,\ 2)$ (2) **Ans.** $(x,\ y)=(3,\ 5)$

3 (1) **Ans.** $(x,\ y)=(-2,\ 1)$ (3) **Ans.** $(x,\ y)=(3,\ -2)$

(2) **Ans.** $(x,\ y)=\left(\dfrac{1}{3},\ -1\right)$ (4) **Ans.** $(x,\ y)=\left(-\dfrac{2}{3},\ -3\right)$

4 Let x be the weight of the jar and y be the weight of sand need to fill the jar.

$$\begin{cases} x+y=100 \\ x+\dfrac{3}{5}y=76 \end{cases}$$

$(x,\ y)=(40,\ 60)$

Ans. 40 grams